Methods and Materials
of Residential Construction

Methods and Materials
of Residential Construction

Laurence E. Reiner, M.E., P.E.

PRENTICE-HALL, INC., ENGLEWOOD CLIFFS, NEW JERSEY   07632

Library of Congress Cataloging in Publication Data

Reiner, Laurence E.
    Methods and materials of residential construction.

    Includes index.
    1. House construction. I. Title.
TH4811.R44   690′.837   81-548
ISBN 0-13-578864-1   AACR2

Editorial/production supervision
    by Ian List and Zita de Schauensee
Interior design by Ian List
Cover design by Edsal Enterprises
Manufacturing buyer: Anthony Caruso

Printed in the United States of America

10 9 8 7 6 5 4 3 2 1

PRENTICE-HALL INTERNATIONAL, INC., *London*
PRENTICE-HALL OF AUSTRALIA PTY. LIMITED, *Sydney*
PRENTICE-HALL OF CANADA, LTD., *Toronto*
PRENTICE-HALL OF INDIA PRIVATE LIMITED, *New Delhi*
PRENTICE-HALL OF JAPAN, INC., *Tokyo*
PRENTICE-HALL OF SOUTHEAST ASIA PTE. LTD., *Singapore*
WHITEHALL BOOKS LIMITED, *Wellington, New Zealand*

contents

v

NINE

STRUCTURAL FRAMING   118

TEN

EXTERIOR WALL CLADDING   163

ELEVEN

ROOFING, FLASHING, GUTTERS, AND DOWNSPOUTS   179

preface

This book is for professional builders; for nonprofessionals who want to do part of their building themselves; and for students, architects, and engineers who would like to learn basic residential construction.

In nonprofessional language it explains the entire process of construction from the proper choice of land to the finishing of the interior ready for occupancy.

By sketches and descriptive language, it illustrates the construction of foundations, the correct way to frame a structure, the interior finishing, and other construction processes.

The author has had more than forty years of experience in this field and has watched and guided both professionals and amateurs during all this time. This book is the essence of his experience.

Darien, Connecticut   Laurence E. Reiner

CHAPTER ONE

selection of the site

One of the most important things that the prospective builder of a home should consider is the character of the area, the neighborhood, and the building lot, and everyone who is planning to build a house should do this before they buy property and build.

1.1 THE OWNER WHO WISHES TO BUILD

To most people the ownership of a house of their own is the largest single investment they will ever make. The choice of the location of this house and the location and character of the building lot on which it is placed can make the difference between becoming a satisfied homeowner or the un-happy occupant of a house. There are people who have built homes on river banks that are subject to periodic flooding or on ocean fronts that are crumbling. There are people who own property in wetlands (which is now subject to very strict building requirements) or on a rock ledge near running water. There is no sewer available and they cannot dispose of their waste and therefore cannot build. There are people who buy or build houses under the approach or takeoff paths of airports or in the path of the wind blowing from a power plant or garbage dump.

All the soot, smell, noise, flooding, and other nuisances can be avoided by a minimum of effort. A few hours spent in driving through an area or examining a zoning map and speaking to a local building official can save a lifetime of frustration and loss of capital investment.

1.1.1   Selection of the Area

People who are looking for building sites come from widely varied social and economic backgrounds, and their choice of locations reflects these differences.

One large market for housing is comprised of young married couples who, after working to build up capital, want a home of their own. It is assumed that most homes are built by currently active wage earners. But there are many cases where people not actively engaged in business on a day-to-day basis, or who have a business of their own, wish to build. Their choice of location may be entirely different.

The choices available to the majority of home builders depend on the size and location of the metropolitan area—big city or large town—in which they work. These choices become fewer in smaller localities, because usually a town expands in only a few directions. This may mean a possible temporary lack of utilities, schools, and other amenities that constitute a favored location.

The City

For the past few years there has been a "back to the city" trend in residential housing. Many people who are tired of driving in heavy traffic or of poor commuting conditions are moving back to the city. There is no question that living in a city with all the amenities it offers is a convenient way of life. Cities offer central-city living conveniences, often at bargain prices, especially if the prospective buyer can do a lot of his or her own repairing and refurbishing. Many cities have set aside older residential areas near the central city as redevelopment areas, and such structures can often be obtained at a very advantageous price. There is a great deal of work to be done in such houses, however, and this does not appeal to many people. The cities also contain semisuburban neighborhoods on their outskirts, and these areas appeal to people who want to live in a city with all its conveniences but still want some trees and lawns around them. Building lots are available in many such areas. In all, there are many choices in city living.

The Suburbs

The suburbs of a city are outside the corporate limits and have a government of their own. They have a separate tax rate, a separate school system, and to a great extent a different mode of life. Many suburbs of large cities have grown so large that they have become cities of their own. The larger suburb is usually closer to the city and with this go higher taxes, mixed school populations, and higher land cost. The farther-out suburbs offer more and lower-priced land and a more relaxed way of life and of course a longer automobile ride or other form of commuting to the city. There are in betweens, of course, and it is suggested that the prospective home builder put down on paper what he or she wants most and concentrate on getting that.

The Country

There are always families or persons who, although attached to a city, want to live in the country. To such people a 2-hour automobile or train ride to the job and back is worth it. The choice of land for a building site in the country requires a great deal of thought and investigation. Country acreage comes as is, with no facilities or utilities or amenities. The land may be reasonably priced, but to this initial cost must be added the cost of a pole line for utilities, a source of water, the building of an access road, the clearing of the land, a sanitary system, and possibly other facilities, such as a gasoline- or diesel-driven electric generator in case of power failure.

1.1.2 Selection of the Community

When the prospective home builder has decided whether he or she wants to live in the city, country, or somewhere in between, he or she must now find a community that is attractive as a place to live and to send children to school, that will provide a comfortable personal environment, and that can be afforded. Every city has suburbs of widely varying character. These range from communities with row after row of small tract houses to strictly zoned small towns which are quite expensive to live in. The home seeker should determine which are possible sites and then investigate each one.

Taxes

Investigate the tax rate. The real estate tax is obviously an important part of any homeowner's budget. The tax rate (the mill rate) is only half the story. The other half is the rate of assessment. This rate can be anywhere from 30% to over 80% of the current market value. A visit to the local tax assessor should give a clear picture of the tax situation. The prospective homeowner should ask a number of questions and jot down the answers. For instance: What is the history of the mill rate for the past several years? What is the rate of assessment? When is the next reassessment due? Is there a special assessment for schools, sewers, or other public improvements? When land is purchased or a house is built, when does the first assessment take place and when is the tax due? (The last question can save many hundreds of dollars in taxes.)

Transportation

Many communities still have a network of public transportation to the nearest major population centers. But in most parts of this country the automobile is the only means of transportation between the home and the job, the food market, and sometimes even the school bus. This makes it difficult for a one-car family, which must locate near a bus route or railroad station and also near a school-bus route. All of this information is available from local real estate agents, the chamber of commerce, or the town hall.

Schools and Amenities

One of the largest migrations of families in recent times has been caused by parents' search for better schools for their children. The level of schooling varies widely in communities. There are some that concentrate on preparing the child for college and others that will give the student a basic education in preparation for the great majority of jobs that do not require a college degree. The second type of school exists in communities where the economic status of the general population is such that the residents are not generally college-oriented. Schools that prepare students for college are likely to have many elective courses in art, literature, and so on, and in consequence a large part of the community taxes go to them. A childless or working couple may wish to avoid such taxes.

Amenities include shops, libraries, theaters, parks, beaches, a golf course, or anything else that appeals to the prospective homeowner. If such amenities are important to the homeowner, they should be considered when choosing a location.

1.1.3   Selection of the Building Lot

How to Find It

When the prospective owner has decided where he or she wants to live—city, suburb, or country and in which particular community—the next step is to find the specific neighborhood and building lot that is most suitable for his or her mode of living and economic status. The building lot may range from a 60- by 100-foot lot in the outlying areas of a city to an acre or more in a suburban or rural area.

Excellent sources of information about land for sale are the advertisements in the Sunday real estate section of the nearest city newspaper. The daily or weekly newspapers of the selected area also advertise land for sale and in addition give the names of local real estate brokers. Purchasing land directly from the owner may save the real estate commission, but it leaves the purchaser without any expert information. Generally, an informed real estate agent really earns the sales commission, which is, in any case, paid by the seller.

The prospective purchaser should first visit the local town hall and speak to a local building official. This is easy to do in a small town or suburb but may be difficult in a city. The writer has found that officials in smaller communities are cordial and quite willing to give the general locations in the community that might meet the purchaser's needs. In a large city, a call to a real estate editor of a local newspaper can be helpful, as is a careful study of real estate advertisments. The writer has even found helpful officials in large cities. It is strongly recommended that the purchaser obtain a zoning map of the area of interest. These are often for sale or can be examined free at the building or the zoning office in the local town hall. Chapter 2 gives a complete explanation of zoning maps and how they should be read. The buyer of land is advised to read the zoning map carefully. There are many cases of homeowners who are completely frustrated or who have to go through lengthy procedures to obtain permission to make changes in their homes because they did not become familiar with the zoning regulations before they bought land or built a house.

Looking for land in rural areas is somewhat more complicated. Newspaper advertisements are always helpful. A visit to the local town hall may or may not produce a zoning map, although today most communities in this country do have some zoning regulations. A detailed map of the area should be available in the town hall and can possibly be purchased. There are companies that specialize in such maps. For instance, the Hagstrom Company,

Inc., 450 W. 33 St., New York, New York 10001, publishes detailed maps of the entire metropolitan area of New York City. The Yellow Pages of telephone directories show map makers and distributors in every major community in the country. A simple telephone inquiry should locate a detailed map of every possible city or suburban area within any metropolitan area. As far as rural areas are concerned, the study of a good road map will show all the primary and secondary roads. The best way to locate land one would like to consider is to drive around and look.

Once armed with information about specific locations and zoning (see Chapter 2), the prospective buyer can approach the owner of the land directly (after finding who this is, by means of a "For Sale" sign, by asking at the nearest house, or by checking tax lists), or the buyer can engage a real estate agent (who, remember, is paid by the seller). The buyer should tell the real estate agent exactly what he or she wants: how much land, the amount of money available, the general area desired, how large a house is planned, details of the family, and any other information that will help the agent to locate an appropriate site.

Topography

The physical features of the building site are very important both from the standpoint of expense and future satisfaction. A piece of flat land that lies just even with or a little below a road may become very soggy during wet weather. A sloping site may drain groundwater into the basement, and wet basements are an everlasting nuisance and expense. Many communities now have wetlands regulations which prohibit any construction within a certain distance of such areas. (See Chapter 2 for a sample of such a regulation.) The prospective purchaser should also be suspicious of large flat areas near a body of water. They could be filled-in swamps or wetlands that were filled before any environmental regulations were in existence. Some of these areas have literally no bottoms, and any structure built on them can be expected to sink and crack, even with the best of foundations. The purchaser should not allow an eager salesperson or developer to talk him or her into buying. If there is any doubt, a town building official can be consulted.

Rocky outcrops should be looked at carefully and suspiciously. Building foundations on ledge rock can be very expensive. If there are nearby houses, the owners may say whether they have encountered any rock. If there are large trees on the property, the buyer should examine their condition. It costs hundreds of dollars to cut up and remove a large tree.

A building site on the top or near the top of a hill can be very attractive. It can also be difficult to reach if the area is subject to snowy and icy roads.

Certain areas of this country have a history of mudslides. This is especially true of land near the Pacific Coast. Most states having such areas have imposed severe restrictions on construction in these areas.

All of the foregoing sounds negative, but there are many people who will still purchase land on river bottoms or filled-in swamps or on a steep rocky hillside. The purpose of these warnings is advisory. The sketches and photographs in Figure 1.1 illustrate some of the topographical problems that can be encountered. Remember that even if all the warnings are heeded, there are still many available building sites.

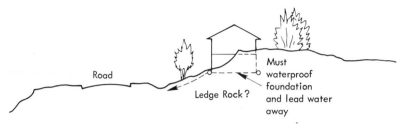

Road

Ledge Rock?

Must waterproof foundation and lead water away

HOUSE ON A HILLSIDE

This house is too near the bluff (which may crumble)

This house may flood

Figure 1.1

You may be buying land that was tidal wetlands not too long ago. This kind of land was formerly a prime target for developers.

These well-laid-out lots are in a ½ acre zone. The house sites assure maximum privacy. Every house in this development was sold before completion.

Flat. Residential zone. Utilities in place. Main line of railroad just beyond rear lot line.

Pretty, woodsy, and wet.

Huge boulders and ledge rock on this commanding site.

Schools

In Section 1.1.2 we discussed the character of schools as they cater to the needs of the locality. But even in the same small city or town, the various schools may each have a different character. This is a choice that must be made before any particular building site in a particular neighborhood is chosen. Inquiries should also be made about school buses and the school to which children from a particular neighborhood will be sent. Parents should be aware that court-ordered busing is now in effect in many areas.

Utilities

The purchaser must know what utilities are in place. If there is no sewer, a septic tank and drainfield will have to be built and maintained. Many small towns are forced by the state government to install a sewer system, and the resultant individual assessment may amount to several thousands of dollars, whether or not the sewer is used. If the building site is a large one and the house is to be set back from the road, the owner may have to pay for a pole line for electricity and telephone service from the street to the house, and many localities now require underground lines. The absence of town water will necessitate an artesian well with pump, pressure tank, and so on. A far-out building site may require an emergency generator in the event of power breakdowns.

This is not to say that the prospective homeowner should not build where he or she pleases, but it is well to be advised that the initial and maintenance costs of such necessities may have to be added to the cost of the land.

Transportation

The transportation that is available from a community may not be within a reasonable distance from the building site. If this necessitates the use of an automobile to reach public transportation, the problem of multiple cars arises again.

1.2   THE BUILDER

In the great majority of cases, the construction of the house will be done by a professional builder. However, the owner may perform part of the work or even act as the general contractor. In any case, the person who does the work should examine the building lot carefully before construction is started.

1.2.1   Planning the House on the Site

By this time the owner has or should have a set of building plans and specifications. Chapter 3 gives the complete details on how a set of plans and specifications tailored to meet the owner's requirement should be prepared. The owner and the builder should now be ready to plan the house on the site.

Limitations of Zoning Regulations

Before the owner and builder even start to locate the position of the house on the building lot, they must be aware of what they are and are not allowed to do by the zoning regulations. Chapter 2 shows how a zoning map should be read and how the regulations affect the location of the house on the lot. Placing the house to obtain a good view, or near a grove of trees, or for other reasons may not be possible because the house would then have to be too near a side line or a front or rear line. Even if within the regulations, such placement might not allow for future expansion. Chapter 2 shows how a building envelope is established.

Slopes, Soil, Rocks, and Sun

The actual building site can be located anywhere within the building envelope. Obviously, the larger the lot, the more leeway the owner will have. Now is the time to locate the house with regard to possible ledge rock, low-lying wetland, trees, and sun direction. On a small lot it may be difficult or even impossible to avoid all these problems. It should be noted, however, that if the owner has not heeded the previous cautionary advice about not purchasing a wet or rocky building site, he or she will have to do the best that is possible using extra land fill or by drainage or even by avoiding altogether a below-grade basement. Sun direction can be a very interesting challenge in both the design and placement of a house. Figure 6.12 shows graphically where the sun rises and sets at different times of the year. If the owner wants sun or shade in various rooms of the house, there is no reason why the location of the rooms within the house cannot be arranged (within cost considerations) to provide this.

Utilities

The location of sewers, water lines, and telephone and electric service lines may also influence the location of the house. The situation and the solutions are different if a well has to be drilled or a sanitary system installed. Various solutions are shown in Chapter 6.

1.3 THE SPECULATIVE BUILDER

The choice of the proper building site is a very important decision for the speculative builder. Whereas an individual owner may indulge personal preferences regarding the type of land he or she builds on, the speculative builder cannot afford to build on land that will not meet the approval of the general public. The site must, of course, conform to the basic needs of any owner, which would include ease of transportation to and from work, the presence of facilities such as shopping, utilities such as water and sewage in place, good schools, and reasonably dry, flat land. In addition to these requirements the speculative builder must consider matters that the ordinary home owner might not think of.

1.3.1 General Requirements for a Speculative Building Site

Growth

The general area must be stable or growing. The actual site should be in the direction of growth in the locality or in a stable residential area. An active builder should be constantly aware of the growing and declining areas and, if financially able, should build up a land bank (see Chapter 5).

Economy of the Area

There are a number of important indicators that will tell the builder whether an area or locality is on the uptrend or at least stable.

Institutions and Government. A state capital or a county seat or a large federal center are sources of underlying stability. So are centers of learning, such as a college or university. These uses attract many satellite uses. At the present time all of these activities are growing and should certainly remain stable in the foreseeable future.

Marketing and Retail. There are many smaller cities and many towns which have become marketing and retail centers for surrounding areas. Such communities present a stable and growing market for new housing.

The Bedroom Community. Every large city and most small ones have satellite suburban areas which serve as bedroom communities. These areas, except for service, have very little business activity of their own. Many of them maintain a pattern of steady growth. In the past few years many central office activities have been moved from cities to these suburban areas. There are great opportunities here for forward-looking builders.

Other Factors. There are other factors to be considered, such as the population trend of the major area, growth of industrial production, growth of retail sales, and so on. These indicators should all be considered when making a decision. Growing or stable communities are where new home purchasers want to be.

1.4 LEGAL MATTERS IN LAND PURCHASE

Property can be transferred from the owner to the purchaser in several ways, depending on local laws and customs. The transfer can be made by an attorney, a title company, a bank, or an escrow officer. In the latter case, the purchaser deposits the money with the escrow officer and the property owner deposits the deed. The sale is consummated when all the conditions of the escrow instrument have been carried out.

The attorney, escrow agent, or other qualified person must first search the title for such items as:

Recorded mortgages or trust deeds.
Easements and rights of way.
Liens against the property.
Restrictions regarding size of building, etc.
Mineral-rights reservations.

The next step is to draw the necessary deed and releases and deliver the deed to the purchaser upon the receipt of the agreed price plus any prorated expenses. The deed must be recorded immediately in the land books of the community. When dealing with a title company, the purchaser also receives a certificate of title, which ensures the purchaser against any defects in the title. In many states the mortgagor will insist upon title insurance as a condition of granting a mortgage.

The owner or builder is warned about purchasing land with encumbrances, such as:

Easements, which give someone other than the owner the right to cross the land by a "right of way," by which someone may at some time build a road to reach other property. The easement might also be for a power or gas line.

Deed restrictions or covenants, which may allow only a certain type or size of house or which contain any restriction on the owner to build what he or she wants (subject, of course, to zoning and building codes).

Riparian rights, which may allow public access to water through the owner's lake-front or beach-front property.

The purchaser is also advised to ask for an abstract of title, which is a condensed version of the title deed history that has been obtained from the legal land records. It summarizes any encumbrances, easements, restrictions, and so on, that go with the land. The abstract should be drawn up by the attorney or title company for the benefit of the purchaser.

1.5 SUMMATION OF LAND CHOICES

It is suggested that the reader fill in Table 1.1 to obtain an overall picture of his or her requirements. There will rarely be a perfect piece of land, but the land that is chosen should be the best compromise. Remember to avoid:

Wetlands.
Filled-in wetland.
Rocky slopes.
River or creek bottoms.
Power stations.
Airports.
Town dumps and incinerators.
Industrial areas.

TABLE 1.1. Summary of Land Choices

	City		City–Suburb		Suburb		Country	
	Yes	No	Yes	No	Yes	No	Yes	No
The general area		√		√	√			
The community								
Ease of transportation						√		
Character of schools					√			
Convenience of shopping						√		
Real estate taxes					√			
Economic status					√			
The building lot								
Neighborhood					√			
Zoning					√			
Character of schools					√			
Utilities						No sewers		
Shopping						√		
Topography					Some rock No wetland			

CHAPTER TWO

zoning and building codes

2.1 THE EFFECT OF ZONING CODES ON RESIDENTIAL CONSTRUCTION

More and more communities in this country are adopting and enforcing zoning regulations to promote the orderly growth of the community. Without an enforced zoning regulation there is nothing to stop the establishment of an automobile junkyard in a residential area or the opening of a bar and grill across the road from a school.

Zoning regulations are often a nuisance to the individual who feels that he or she should be able to do whatever he or she wishes on privately owned land, but such regulations are established and enforced to promote the general welfare. It has been found that in communities with reasonably restrictive and well-enforced zoning regulations, the property values have been stable or rising and neighborhoods have been stable. Lending institutions favor such areas and make it easier to obtain mortgage money.

In the past few years, as the country has become more ecology-conscious, the zoning boards of many communities have taken on the added duties of conserving ecological resources. The purpose of one such regulation follows:

To conserve and protect all of the natural resources of the Town, including land, soil, air, water, wetlands, marshes, ponds and lakes, streams and water courses, shore-front and coastal lands, rivers and tidal estuaries, trees and vegetation, forests, aquifers and water tables, wildlife, areas of scenic beauty, and areas of ecological importance—in recognition of the important interrelationships between these resources and a suitable environment for human habitation, their direct influence on the Town's suitability for residential use, and their importance to the health, safety and general welfare of the Town and its larger environs.

The person who wishes to buy land on which to build any residential structure must therefore be familiar with the general purposes of zoning and must particularly become familiar with the zoning regulations of the community where he or she wishes to build. Careful study of the regulations will reveal what can be built and where. Zoning regulations are somewhat difficult to interpret, however, and the following section explains them by giving examples of their use.

2.2   EXAMPLES OF TYPICAL ZONING REGULATIONS

Every zoning regulation has an accompanying map of the area that is covered by that regulation. The map is prepared on a street-by-street basis, and various areas are marked, usually by letters and numbers, to show what kind of activity can be carried on within the boundaries of the map. Figure 2.1 shows a zoning map. The zoning regulation will contain a table or schedule that explains what each zone designation means. In this case the one-family residential area identification is shown on the map. This particular town does not allow any multiple dwellings whatsoever, which is not usually the case.

It will be noted that the zones that call for larger land areas [R-1 (1 acre) and R-2 (2 acres)] are the farthest removed from business areas. As the land areas become smaller, such as R-$\frac{1}{3}$ and R-$\frac{1}{5}$, they become immediately adjacent to such areas.

The prospective home builder looking for land should examine the zoning map carefully and try to be as far away from undesirable zones as possible. One should also be aware that not all business zones are alike. In this case the town does not allow heavy industry or other nuisance uses, but there are still allowed uses that are best to avoid. For instance:

PB (Planned Retail Business)

Allows retail stores, professional and business offices, banks and post offices, restaurants, and cleaning and dyeing businesses, but with provision that all service trucks must be parked in a permanent structure. It also allows parking for passenger cars only. These are reasonably limited uses with little nuisance effect.

CB (Central Business)

This zone is more liberal in its allowed uses and in addition to the retail shops and the like in the PB zone, it allows, among other uses, clinics, clubs and lodges, community center, garden supplies, laundries, pet shops, taxi stands, and theaters. These uses will certainly create more noise and traffic, as well as more use of parking, than will the PB zone.

SB (Service Business)

The allowed uses include auto sales agencies, bowling alleys, gasoline service stations, garages, secondhand and auction stores, cat and dog hospitals, and other uses that are unsightly and that create noise and possibly unpleasant odors.

There are other zones, but SB is the one that is best to stay away from.

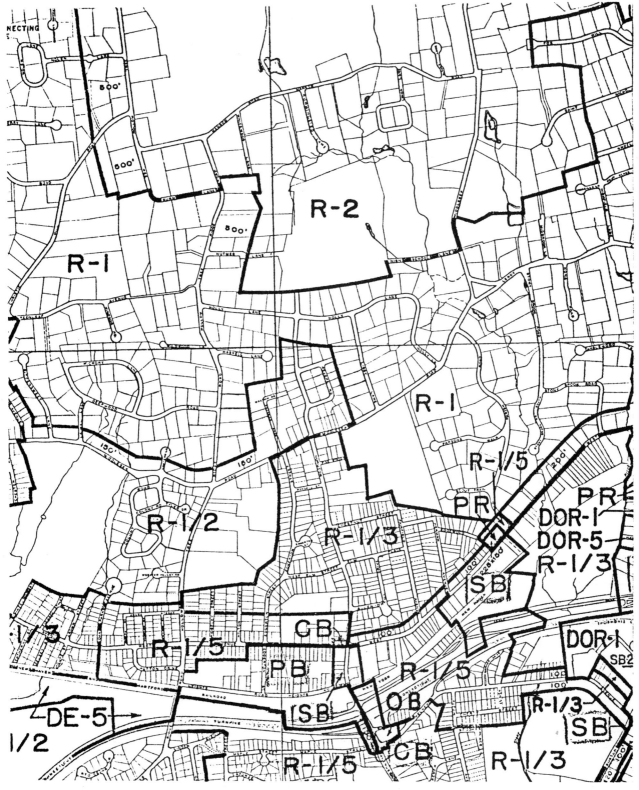

R–2:   One-family residential, 2 acres
R–1:   One-family residential, 1 acre
R-1/2: One-family residential, $\frac{1}{2}$ acre
R-1/3: One-family residential, $\frac{1}{3}$ acre
R-1/5: One-family residential, $\frac{1}{5}$ acre

Figure 2.1 Section of a typical zoning map.

15

Figure 2.2. Section of a big city zoning map.

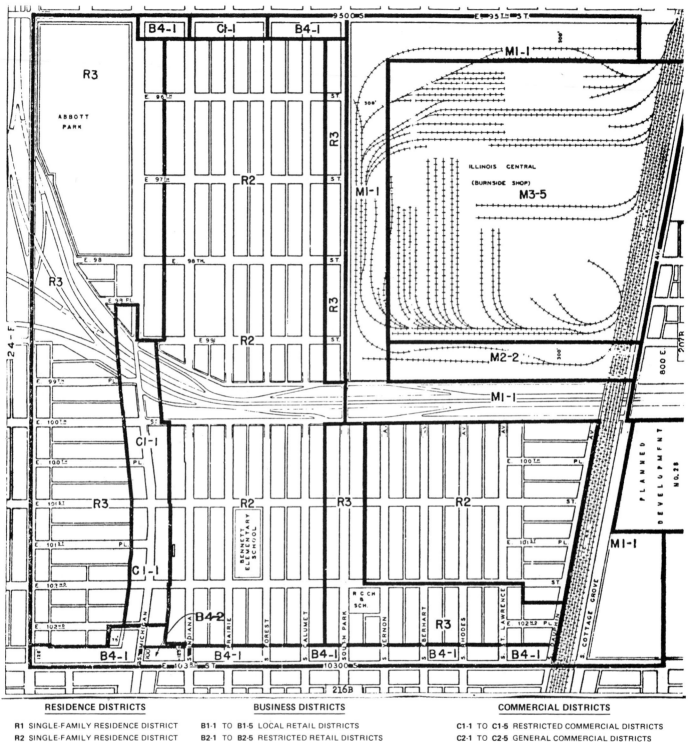

RESIDENCE DISTRICTS

R1 SINGLE-FAMILY RESIDENCE DISTRICT
R2 SINGLE-FAMILY RESIDENCE DISTRICT
R3 GENERAL RESIDENCE DISTRICT
R4 GENERAL RESIDENCE DISTRICT
R5 GENERAL RESIDENCE DISTRICT
R6 GENERAL RESIDENCE DISTRICT
R7 GENERAL RESIDENCE DISTRICT
R8 GENERAL RESIDENCE DISTRICT

BUSINESS DISTRICTS

B1-1 TO B1-5 LOCAL RETAIL DISTRICTS
B2-1 TO B2-5 RESTRICTED RETAIL DISTRICTS
B3-1 TO B3-5 GENERAL RETAIL DISTRICTS
B4-1 TO B4-5 RESTRICTED SERVICE DISTRICTS
B5-1 TO B5-5 GENERAL SERVICE DISTRICTS
B6-6 AND B6-7 RESTRICTED CENTRAL BUSINESS DISTRICTS
B7-5 TO B7-7 GENERAL CENTRAL BUSINESS DISTRICTS

COMMERCIAL DISTRICTS

C1-1 TO C1-5 RESTRICTED COMMERCIAL DISTRICTS
C2-1 TO C2-5 GENERAL COMMERCIAL DISTRICTS
C3-5 TO C3-7 COMMERCIAL-MANUFACTURING DISTRICTS
C4 MOTOR FREIGHT TERMINAL DISTRICT

MANUFACTURING DISTRICTS

M1-1 TO M1-5 RESTRICTED MANUFACTURING DISTRICTS
M2-1 TO M2-5 GENERAL MANUFACTURING DISTRICTS
M3-1 TO M3-5 HEAVY MANUFACTURING DISTRICT

500'   250'   0   250'   500'   750'   1000'

SCALE IN FEET

16

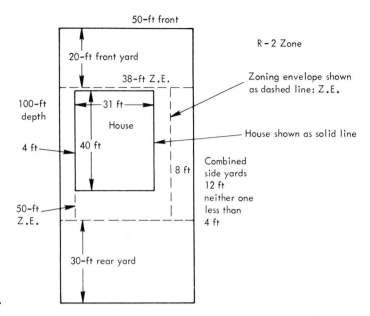

Figure 2.3. Big city zoning envelope.

In the larger towns and cities, there are allowed uses that include heavy, noisy, and odorous industry.

In this case, unlike in smaller towns, the lower-zoned residential areas allow multiple dwellings and may be near medium or heavy industrial zones. The advice to the would-be purchaser of land is to look at a zoning map before you buy. It may be a pleasant tree-lined street downwind from a fat-rendering plant. If there is any question in a purchaser's mind, he or she is strongly advised to consult a zoning official. Such officials are generally helpful and will point out possible nuisance areas. It may involve a trip to the city hall, but it is worth it to avoid a possible costly mistake.

The zoning map shown in Figure 2.2 is a portion of the zoning map of a large city. The residential areas shown here are R–2 and R–3. As shown in Figure 2.3, the side, front, and rear yard requirements in R–2 allow a zoning envelope 38 ft wide by 50 ft deep on a minimum lot area of 5000 ft^2. In this R–2 zone the maximum floor area ratio is 0.5. This means that the total area of the building cannot be more than 0.5 ($\frac{1}{2}$) of the area of the land. This means that no more than 2500 ft^2 of house can be built. If it is a two-story residence, it can contain 1250 ft^2 per floor. This is roughly 31 ft by 40 ft or any similar combination. Placing such a house on the lot as shown gives the owner some planting and recreational area.

This would seem like a good area in which to build a modest house within a city limit, with all of its consequent conveniences. It should be noted, however, that there is a large railroad yard and railroad shops nearby. There is also a railroad right of way. In this case the prospective owner would be well advised to ask present residents about the problem of noise and to spend some time to determine this personally.

2.3   HOW TO INTERPRET A ZONING REGULATION

In many aspects of zoning, there have been numerous instances of home-owners who are unable to alter or enlarge their houses because insufficient attention was paid to the zoning regulations before the house was built.

2.3.1   The Zoning Envelope

Following are a number of diagrams to show how a zoning envelope should be laid out to allow some leeway in future expansion. It must be understood, of course, that if the owner starts with the largest possible house allowed in that zone, nothing can be added later. But the usual zoning envelope is so large that this possibility is somewhat remote.

Table 2.1 shows allowed areas and house coverage in the community shown in Figure 2.1. For the purpose of this book, each building lot will be shown with parallel lines and right angles. In the following illustrations, the writer has not used odd-shaped lots (which will be discussed in Section 2.3.2). There is nothing wrong with such lots as long as they meet the minimum frontage, width, depth, and area requirements.

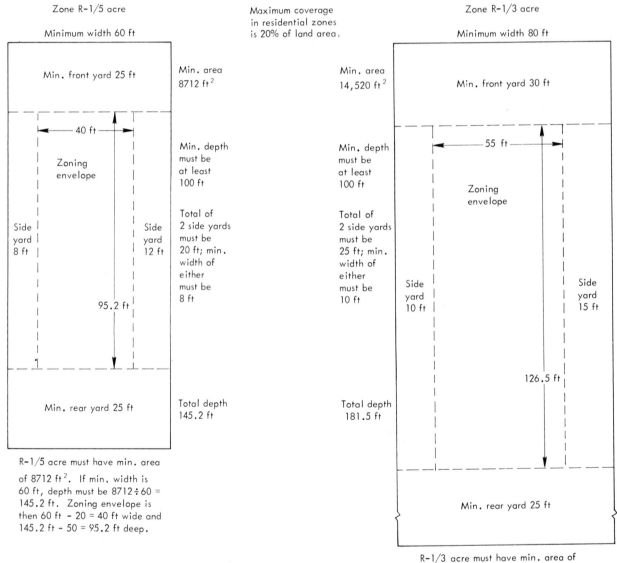

Zone R-1/5 acre

Minimum width 60 ft

Min. front yard 25 ft

Min. area 8712 ft^2

40 ft

Zoning envelope

Min. depth must be at least 100 ft

Total of 2 side yards must be 20 ft; min. width of either must be 8 ft

Side yard 8 ft

Side yard 12 ft

95.2 ft

Min. rear yard 25 ft

Total depth 145.2 ft

R-1/5 acre must have min. area of 8712 ft^2. If min. width is 60 ft, depth must be 8712 ÷ 60 = 145.2 ft. Zoning envelope is then 60 ft - 20 = 40 ft wide and 145.2 ft - 50 = 95.2 ft deep.

Maximum coverage in residential zones is 20% of land area.

Min. area 14,520 ft^2

Zone R-1/3 acre

Minimum width 80 ft

Min. front yard 30 ft

Min. depth must be at least 100 ft

Total of 2 side yards must be 25 ft; min. width of either must be 10 ft

55 ft

Zoning envelope

Side yard 10 ft

Side yard 15 ft

126.5 ft

Total depth 181.5 ft

Min. rear yard 25 ft

R-1/3 acre must have min. area of 14,520 ft^2. If min. width is 80 ft, depth must be 14,520 ÷ 80 = 181.5 ft. Zoning envelope is then 80 ft - 25 = 55 ft wide and 181.5 ft - 55 = 126.5 ft deep.

Figure 2.4. Typical zoning envelopes.

TABLE 2.1. Allowed Areas and House Coverage

	R-2	R-1	R-$\frac{1}{2}$	R-$\frac{1}{3}$	R-$\frac{1}{5}$
Minimum lot area (ft)^2	87,120	43,560	21,780	14,520	8,712
Minimum width (ft)	200	150	100	80	60
Minimum frontage (ft)	75	50	50	50	50
Minimum depth (ft)	200	150	100	100	100
Minimum front yard (ft)	50	40	40	30	25
Minimum side yard (ft)					
Least one	35	25	15	10	8
Total of two	70	50	30	25	20
Minimum rear yard (ft)	50	40	30	25	25
Accessory buildings (ft)					
Minimum distance from side lot line	35	25	10	5	5
Minimum distance from rear lot line	50	40	10	5	5
Maximum height (stories)	$2\frac{1}{2}$ (all residential zones)				
Maximum height (ft)	30 (all residential zones)				
Maximum building coverage	20% (all residential zones)				
Minimum number of off-street parking spaces	Two for each family unit (all residential zones)				

Figure 2.4 shows typical zoning envelopes. The structure can be located anywhere within the zoning envelope. This shows that there is room for expansion on even the smallest building lot if some forethought is given to the location of the structure. There are many communities that allow a 50-by 100-ft lot. In such cases the legal side yards and the front and rear yards are smaller, so that there will be room for some expansion.

2.3.2 Odd-Shaped Lots

It will be noted in Table 2.1 that the minimum allowed frontage in the various zones is less than the minimum allowed width. This seems like a contradiction, but it is done for a definite purpose, because not every building lot is a perfect oblong with parallel sides. The illustration shown here (Figure 2.5) is an example of an odd-shaped lot in zone R-$\frac{1}{5}$. The frontage is 50 ft, but it widens to the allowed minimum of 60 ft at a distance of 40 ft from the front. The regulation states that no structure can be built in any portion of the lot that is less than 60 ft wide. This means that the house must be built at least 40 ft back from the front. As shown here, there is plenty of room to build a house within the zoning envelope and still preserve the side and rear yard requirements.

2.3.3 How to Locate the House on the Lot

We have seen that the area that is allowed for a residential structure, even in a most restrictive zoning regulation, is more than sufficient to enable the owner to build a good-size house. Even in the smallest allowable lot, a house of 40 ft by 30 ft by $2\frac{1}{2}$ stories will provide at least 3000 ft^2 of livable area (including a garage), plus a basement. However, in placing the house on the lot, the owner must think ahead. Using the R-$\frac{1}{5}$ zone as an example, Figure 2.6 shows that if the house is placed directly on the front

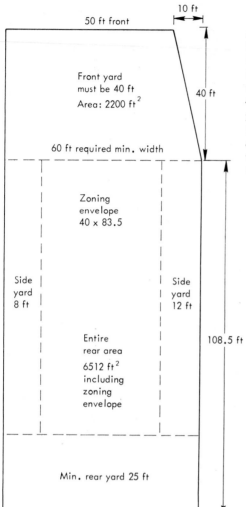

R-1/5 acre — 8712 ft^2

10 ft

50 ft front

Front yard
must be 40 ft
Area: 2200 ft^2

40 ft

60 ft required min. width

Zoning
envelope
40 x 83.5

Side
yard
8 ft

Side
yard
12 ft

108.5 ft

Entire
rear area
6512 ft^2
including
zoning
envelope

Min. rear yard 25 ft

Figure 2.6. Placing the house on the lot, zone R-1/5.

Area of the front yard to the allowed building line is 60 x 40, or 2400 ft^2, minus the area of triangle:
40 x 10 ÷ 2 = 200 ft^2 or 2200 ft^2.
The rear area is 6512 ft^2: 6512 ÷ 60 = 108.5 ft, which is the depth of the rear area.

Figure 2.5. Odd-shaped lot.

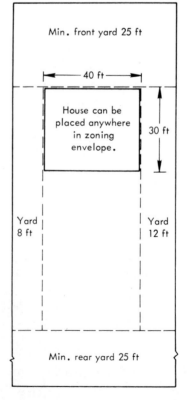

Min. front yard 25 ft

40 ft

House can be
placed anywhere
in zoning
envelope.

30 ft

Yard
8 ft

Yard
12 ft

Min. rear yard 25 ft

Maximum coverage in this zone is 20% or 1742 ft^2. A house can be 40 by 30 ft, which could include a 10-ft-wide garage facing the street. Such a house can include a living room, a dining space, a kitchen, a powder room, and a rear porch on the first floor. There can be 3 bedrooms and a bath on the second floor. There can also be an attic and a basement. The 40- by 30-ft site shown here can be changed to any dimensions as long as the total area of the building does not exceed the allowed coverage.

20

line, nothing further can be built in front, but the house can be extended for another 64.2 ft in the rear. The same principle applies to the side and rear yards. If the structure is built up to the zoning limits, nothing further can be added in these areas.

2.3.4 Filing of Original Plot Plan and Filing an Appeal

The application for a building permit has to be accompanied by a site or plot plan, which shows the size of the lot, the size of the house, and other information. This application is for a particularly strict enforcement area, but if the prospective owner or builder studies this one carefully, he or she should be prepared for most others.

If for some reason the owner wishes to build a house that does not strictly comply with the zoning regulations or if for some unforeseen reason he or she has to add to the house and thereby violate a side or front or rear line or cover more than the allowed area, he or she must file an appeal. When such an appeal is made, the neighboring property owners must be notified so that they can object, if they wish, to any violation of the regulations.

It is difficult to obtain a variance of the zoning, and usually such appeals are not granted.

2.4 BUILDING CODES FOR RESIDENTIAL PROPERTY

When the owner and builder have satisfied themselves that the proposed structure will fit within the zoning envelope, the next step is to be sure that the plans and specifications meet the requirements of the local building code.

Every locality in the country maintains and enforces a building code. The purpose of this code is to provide minimum standards of construction "for the protection of public health, welfare, and safety." The code will specify the requirements for the strength of the foundations, the structural frame, the exit requirements for fire safety; the standards for sanitary facilities, light, and ventilation; and many other related matters.

Building codes are written to regulate every kind and type of construction. However, the builder and owner of residential property need only become familiar with the relatively small part of the code specific to his or her situation.

There are a number of building codes used in this country. There are several national codes that have been adapted to their own use by many states. Such statewide codes are then used by all the communities in the state, with the possible exception of the larger cities, which have changed the code to fit their requirements. However, the basic provisions as set forth here are typical for any code anywhere. The code used here is one that has been adopted by many states—BOCA—Building Officials and Code Administrators International, 1313 East 60 St., Chicago, Illinois 60637. To find the code under which any area operates, one should call a local building official.

If the reader will keep these basic requirements in mind, it should be a simple task to inquire at any community building office to ascertain in what way their requirements differ. *It should be noted that the following code provisions are typical only and can be changed (and are) by local action. The builder must check each individual local code.*

2.4.1   Administration of the Code

The provisions of a code apply to the construction, alteration, addition, repair, removal, demolition, use, location, occupancy, and maintenance of all buildings and structures, whether existing or proposed within the jurisdiction of the code. This is an all-embracing statement and is in fact strictly enforced in most localities. The owner and builder must therefore be sure that the plans for the new construction, whether they be for alteration, addition, or a new structure, are examined and approved by the local authority. This usually involves filing the plans at the local building office. In many communities the building office is combined with the zoning office, and this means that a plot plan must also be filed, to show that the construction will comply with the zoning regulation.

Although every community enforces a building code, there are degrees of enforcement. The best way to find how strictly the particular locality enforces its code is to speak to the local building official. The official should be told what is planned and his or her advice should be obtained as to how to go about obtaining a building permit or whatever else is required before construction can proceed. A copy of a typical building permit form is shown in Figure 2.7.

The owner and builder should expect that the building inspector will come at any time to check that the construction meets code requirements. As a matter of fact, the builder must call the building inspector at certain stages of construction so that the inspector can be assured that the construction meets such requirements. Such stages might include: before foundation footings are poured; when the structural frame is erected; before the plumbing or electric work is covered; and when the construction is complete (Figure 2.8). At completion, the building official issues a certificate of occupancy. This is a legal document which shows the lending institution or any mortgagee that they may now make their final payment on the construction loan.

2.4.2   Area and Height Limitations

The usual one- or two-family structure is of frame construction and is defined as follows: A structure in which the exterior walls, bearing walls, partitions, floor, and roof construction are constructed wholly or partly of wood stud and joist assemblies with a minimum nominal dimension of 2 in. or of other approved combustible materials with firestopping at all vertical and horizontal draft openings. Such a structure is not allowed to be more than 35 ft or $2\frac{1}{2}$ stories high and may not have a floor area of more than 4800 ft^2 per floor. If greater height or more area per floor is required for a multifamily structure, the structural elements must be changed. For instance, a "protected" frame residential structure may be a full three stories high and may have an area of 10,200 ft^2 per floor. "Protected" construction usually means exterior masonry walls, fire-rated interior partitions, masonry fire walls, fire-rated exitways, and fire-treated wood structural members.

APPLICATION FOR BUILDING PERMIT

The undersigned owner or authorized agent hereby applies for a permit to construct a building in accordance with the laws and the ordinances of the State of   and the Town of   and as set forth in the accompanying drawings and specifications insofar as the same shall be found not to conflict with the aforesaid State and Town laws, and also for a Certificate of Occupancy for the use as herein stated.

Date _____ 19_____

House Number _____ Lot Number _____ Street _____

Owner _____

Owner's Address _____
                          Street   City   State

Please check items below that apply to you

CONSTRUCTION: New ☐   Alteration ☐   Addition ☐   Repair ☐   Removal ☐

TYPE OF OCCUPANCY: One Family Residence ☐   Garage ☐   Shed ☐   Pool ☐
   Other _____

FOUNDATION: Basement Yes ☐   No ☐ Walls: Poured Concrete ☐ Blocks ☐ Other _____
   Mason's name _____ Address _____

STRUCTURE: Frame ☐   Brick ☐   Stone ☐ Conc. Block ☐   Other _____
   Carpenter's name _____ Address _____

PLUMBING: Connect to Sanitary Sewer ☐   Septic Tank ☐
   Connect to City Water ☐ Other water supply _____
   Number of Fixtures: Bath tubs _____ Sinks _____ W.C. _____ Lavatories _____ Laundry tubs _____
   Plumber's name_____ Permit No. _____

HEATING: Heat by Electric ☐   Oil ☐   Gas ☐   Hot air ☐   Hot water ☐   Steam ☐

ELECTRICAL WORK must conform with the National Electrical Code
   Electrician's name _____ Permit No. _____
   Oil Burner Installer's name _____ Permit No. _____

Please answer the following questions:

Size of Building _____ Number of Floors _____ Floor Area _____ Zone _____

Size of Addition _____ Size of Pool _____

Dimensions of Lot _____ feet front, by _____ feet deep _____

Is there a building on this lot now _____ If so, how occupied _____

Architect's name _____ Connecticut Reg. No. _____

General Contractor's name _____ Address _____

Be sure to fill in Data on inside of application

   I estimate the value of this work will be $_____

   Applicant's Signature _____

   Address _____

   City _____ Phone _____

Date Issued | Permit Number

Figure 2.7. Application for building permit.

DO NOT REMOVE

BUILDING DEPARTMENT

Permit No. _____

Has been issued to _____

Address _____

To erect a _____

No inspections will be made unless this card is prominently displayed on front of building, properly protected from the weather, and easily accessible to the Inspectors.

Inspections to be signed by authorized Inspectors

Soil For Footing   Date   FORMS IN PLACE READY FOR
                                        CONCRETE
                                Insp. by
Pour no concrete until above has been signed _____

Foundation before backfilling _____

Location Survey
No additional inspections until received by Bldg. Dept. _____

Plyscore Inspection
For nailing and grading _____

Rough Electrical _____

Rough Plumbing _____

Framing
Do not insulate until the above have been approved _____

Burner Permit No. _____

Final Inspection: _____

Electrical _____

Plumbing _____

Building _____

Health Insp. _____

Fire Marshal _____

Figure 2.8. Record of inspections.

2.4.3   Foundations

The code specifies the kind of underlying material that is satisfactory for foundations. Table 2.2 shows the bearing value of each type of soil or rock.

There may be instances where the building official is not satisfied that the underlying soil will bear the weight of the structure. This may be especially true when the land is located on a steep slope or in soggy ground or on suspected land fill. In such a case, the following excerpt regarding foundation investigations from a code will show what may have to be done.

TABLE 2.2. Presumptive Surface Bearing Values of Foundation Materials[a]

Class of Material	Tons per Square Foot
1. Massive crystalline bedrock, including granite, diorite, gneiss, trap rock, hard limestone, and dolomite	100
2. Foliated rock, including bedded limestone, schist and slate in sound condition	40
3. Sedimentary rock, including hard shales, sandstones, and thoroughly cemented conglomerates	25
4. Soft or broken bedrock, (excluding shale), and soft limestone	10
5. Compacted, partially cemented gravels, and sand and hardpan overlying rock	10
6. Gravel and sand-gravel mixtures	6
7. Loose gravel, hard dry clay, compact coarse sand, and soft shales	4
8. Loose, coarse sand and sand-gravel mixtures and compact fine sand (confined)	3
9. Loose medium sand (confined), stiff clay	2
10. Soft broken shale, soft clay	1.5

[a]Lightweight structures: Mud, organic silt, or unprepared fill shall be assumed not to have presumptive bearing capacity unless approved by test, except where the bearing capacity is deemed adequate by the building official for the support of lightweight and temporary structures.

When required: In the absence of satisfactory data from immediately adjacent areas, the owner or applicant shall make borings, test pits, or other soil investigations at such locations and to sufficient depths of the bearing materials to the satisfaction of the building official. For all buildings which are more than three (3) stories or forty (40) feet in height, and whenever it is proposed to use float, mat or any type of deep foundation, there shall be at least one (1) exploratory boring to rock or to an adequate depth below the load-bearing strata for every twenty-five hundred (2500) square feet of built-over area, and such additional tests that the building official may direct. When the safe sustaining power of the soil is in doubt, or superior bearing value than specified in this code is claimed, the building official shall direct that the necessary borings or tests be made.

Other provisions of a typical code are:

Foundation footings must extend below the frost line.

Footings cannot be poured or laid on frozen ground.

Footings must project at least 4 in. on either size of the foundation wall and must be at least 8 in. thick.

Footings must be of sufficient width and thickness to spread the load according to the bearing capacity of the soil.

Foundation walls may be of solid or hollow masonry or of poured concrete.

The top course of hollow masonry blocks must be filled solidly with mortar, and anchor bolts ½ in. in diameter by 15 in. long should be inserted at proper intervals. Anchor bolts inserted in concrete walls need be only 8 in. long. At least two such bolts are required for each section of plate. (Many communities do not require anchor bolts.)

Foundation walls must be of at least the same width as the walls they support.

Foundation walls of less than 6 ft in height may be 8 in. wide. If over 6 ft high, they must be 12 in. wide.

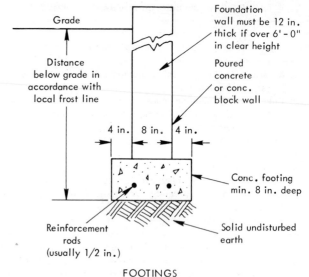

Grade

Distance
below grade in
accordance with
local frost line

Foundation
wall must be 12 in.
thick if over 6'-0"
in clear height

Poured
concrete
or conc.
block wall

4 in.   8 in.   4 in.

Conc. footing
min. 8 in. deep

Reinforcement
rods
(usually 1/2 in.)

Solid undisturbed
earth

FOOTINGS

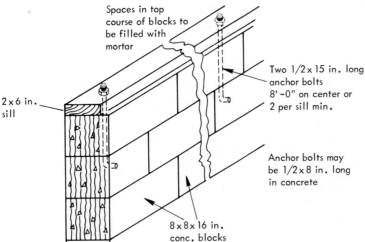

Spaces in top
course of blocks to
be filled with
mortar

2 x 6 in.
sill

Two 1/2 x 15 in. long
anchor bolts
8'-0" on center or
2 per sill min.

Anchor bolts may
be 1/2 x 8 in. long
in concrete

8 x 8 x 16 in.
conc. blocks

FOUNDATION WALLS

Figure 2.9. Some illustrations of code requirements (foundations).

Some of these requirements are illustrated in Figure 2.9.

It is very rarely necessary in the case of two-family residences to construct other than poured concrete or masonry footings. But in some areas that have unstable underlying soil, it may become necessary to drive some kind of pile foundations. The code names the types of piling that can be used and what its bearing capacity should be.

2.4.4 Structural Framing

The foreword of the code usually states that the structure should be designed and constructed to be of sufficient strength and rigidity to safely support all loads, including dead loads. The dead load is the weight of the permanent structural and nonstructural components of a building, such as walls, floors, and fixed service equipment. The live load is the weight superimposed by the use and occupancy of the building, such as people, furniture, and movable equipment.

TABLE 2.3. Code Requirements for Allowable Maximum Spans

Size (ft)	Spacing (in.)	Maximum Clear Span (ft–in.)
Floor Joists		
2 × 6	12	10–3
	16	9–4
	24	8–2
2 × 8	12	13–6
	16	12–3
	24	10–9
2 × 10	12	17–3
	16	15–8
	24	13–8
2 × 12	12	21–0
	16	19–1
	24	16–8
Rafters		
2 × 6	12	12–6
	16	10–10
	24	8–10
2 × 8	12	16–6
	16	14–4
	24	11–8
2 × 10	12	21–1
	16	18–3
	24	14–11

Allowable live loads for residential structures are:

In Multifamily Houses

Private apartments	40 lb/ft²
Public rooms	100 lb/ft²
Corridors	80 lb/ft²

In Private Dwellings

First floor	40 lb/ft²
Second floor	30 lb/ft²
Habitable attic	30 lb/ft²
Uninhabitable attic	20 lb/ft²

The sizes and spacing of joists, girders, studs, rafters, and any other structural lumber are standard everywhere. Table 2.3 is a portion of the tables published by the National Forest Products Association and shows such typical sizes.

Other provisions of the code are as follows:

Joists must bear on at least one-fourth of their depth.

Corners of the structure must consist of three 2 x 4's and must be corner-braced with 1 x 4 diagonal braces unless diagonal wood sheathing or plywood sheathing is used.

Lintels over all openings must be of sufficient strength to transfer their loads equally to the supporting members.

Firestopping must be provided in stud walls and partitions between each floor.

The ceiling joists that meet the roof rafters must be securely fastened to the rafters.

The rafters must be either vertically supported at the ridge or must be tied together with 1 x 6 collar beams at at least 5-ft intervals and at a height equal to two-thirds of the distance from the finished floor to the peak.

Exterior Weather Boarding, Veneers and Condensation: To secure weather tightness in framed walls and other unoccupied spaces, the exterior walls shall be faced with an approved weather-resisting covering properly attached to resist wind and rain. The cellular spaces shall be so ventilated as not to vitiate the firestopping at roof, attic and roof levels or shall be provided with interior non-corrodible vapor-type barriers complying with the approved rules; or other means shall be used to avoid condensation and leakage of moisture. The following materials shall be acceptable as approved weather coverings of the nominal thickness specified:

Brick masonry veneers	2 in.
Stone veneers	2 in.
Clay tile veneers	$\frac{1}{4}$ to 1 in.
Stucco or exterior plaster	$\frac{3}{4}$ in.
Precast stone facing	$\frac{5}{8}$ in.
Wood siding (without sheathing)	$\frac{5}{8}$ in.
Wood siding (with sheathing)	$\frac{1}{2}$ in.
Protected fiber board siding	$\frac{1}{2}$ in.
Wood shingles	$\frac{3}{8}$ in.
Exterior plywood (with sheathing)	$\frac{5}{16}$ in.
Asbestos shingles	$\frac{5}{32}$ in.
Asbestos cement boards	$\frac{1}{8}$ in.
Aluminum clapboard siding	0.024 in.
Formed steel siding	29 gage
Hardboard siding	$\frac{1}{4}$ in.

Figure 2.10 illustrates some of these provisions.

Modulus of Elasticity

In interpreting a code or any other stated requirement for structural strength, the builder will often come across the letter E or *modulus of elasticity*. This denotes the strength of a structural member. The builder need only know what the E for any grade of lumber is and know that the

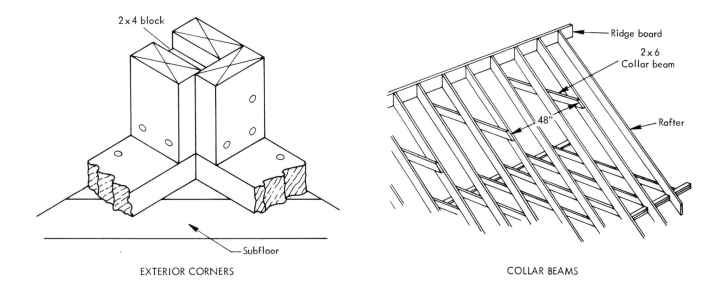

2 x 4 block

Ridge board

2 x 6
Collar beam

48"

Rafter

Subfloor

EXTERIOR CORNERS

COLLAR BEAMS

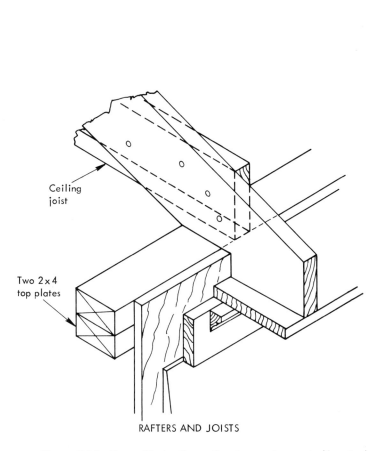

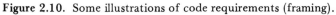

Ceiling
joist

Two 2 x 4
top plates

RAFTERS AND JOISTS

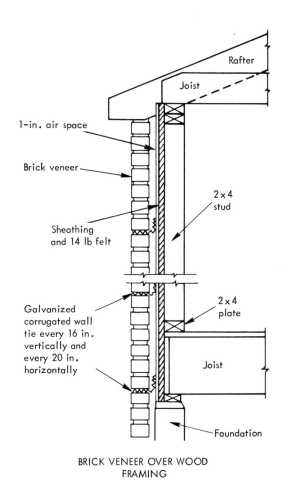

Rafter

Joist

1-in. air space

Brick veneer

2 x 4
stud

Sheathing
and 14 lb felt

2 x 4
plate

Galvanized
corrugated wall
tie every 16 in.
vertically and
every 20 in.
horizontally

Joist

Foundation

BRICK VENEER OVER WOOD
FRAMING

Figure 2.10. Some illustrations of code requirements (framing).

29

higher the E, the better and stronger the lumber. For instance, the E for Douglas fir and southern pine is 1,760,000, whereas the E for eastern hemlock or fir is 1,210,000. The builder who uses these weaker woods for structural members must use more and heavier members to come up to code requirements. It is a matter of cost and future maintenance.

Table 2.3 is based on the use of common structural-grade Douglas fir or No. 1 common southern pine, whose modulus of elasticity is 1,760,000. However, to allow for errors in grading, the numbers shown here are based on a modulus of elasticity of only 1,400,000. If lower-strength lumber is used, these figures must be modified.

2.4.5 Masonry

Some of the basic provisions of the code for masonry work are as follows:

Masonry veneer walls over wood frame construction must be securely fastened to the wood frame with corrosion resistance ties at 16-in. vertical and 20-in. horizontal spacing and must be backed with 14-lb felt paper. Veneer walls 2 in. thick cannot exceed 18 ft above their support and veneer walls 4 in. thick cannot exceed 25 ft above their support.

Masonry bearing walls must have a header course or a heavy corrosion-resistant corrugated wall tie at every sixth course. There must be at least one wall tie every $1\frac{1}{2}$ ft^2.

Masonry walls 35 ft high (including subgrade) must be 8 in. thick.

Cavity walls must be at least 10 in. thick up to 25 ft in height and at least 14 in. thick up to 40 ft in height. Neither of the courses of a cavity wall can be less than 4 in. thick.

All masonry walls must be supported at right angles at intervals of at least 14 times their thickness.

2.4.6 Light and Ventilation

The purpose of this section of a building code is to require the builder to provide the minimum of light and ventilation necessary for the health, safety, and welfare of the occupants. For instance:

Windows and exterior doors may be used for light and ventilation. In such case their aggregate glass area must be at least one-tenth of the floor area and one-half of this area must be available for ventilation.

Example: 10- $\times$ 12-ft room = 120 ft^2

  Glass area at least 12 ft^2

  Openable area at least 6 ft^2

Such windows or doors must open directly on a street, a yard, or an alley on the same lot.

Mechanical ventilation may be used, but there must be at least two full changes of air per hour and such system must be in operation at all times that it is occupied.

Bath and toilet rooms must have a window or a skylight of at least 3 ft², part of which can be opened for ventilation.

A basement cannot be used as a habitable room unless at least one-half of its height is above grade.

Any natural light must provide at least 6 foot-candles of illumination at a height of 30 in. above the floor level.

2.4.7   Fire Protection and Means of Egress

The requirements for the fire protection of residential property are quite simple for single-family residences.

Firestopping is required to close all concealed draft openings such as plumbing stacks between floors and to close all framing openings between floors.

Wood framing cannot be supported by the masonry used for fireplace or furnace chimneys but must be framed around it (Figure 2.11).

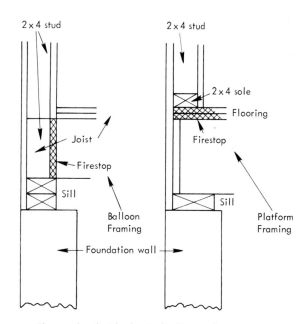

Figure 2.11. Code requirement for firestopping.

The crosshatched lumber is the firestop that prevents a fire from spreading up between the vertical studs

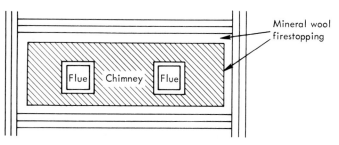

Framing and firestopping around a chimney

In the case of a two-family house: when one family lives above the other, the floor between them must be of 1-hr fire resistance. When families live side by side, such a fire wall must be built between each family unit.

Stairways in single- or two-family residences may be of wood construction but must be at least of 3 ft clear width. Stairways cannot rise more than 12 ft between landings. The treads must be at least 9 in. wide and the risers not more than 8¼ in. high.

In a multiple dwelling a means of exit (a fire-protected stairway) can be no more than 100 ft from the farthest point of a dwelling unit.

Fire walls between dwelling units must be of 1-hr rating.

2.4.8   Plumbing

The primary objective of the plumbing code is in the protection of the public health. To attain this there are certain basic rules:

Every dwelling unit must have a supply of pure and wholesome potable water, and such water supply must not in any way be cross-connected to or in any way be capable of being polluted by waste or sewage or unsafe water of any kind.

Wastes must be disposed of by connection to a sewer or, in its absence, to an approved sewage disposal system.

Every dwelling unit must have a sink, a tub or shower, a water closet, and a basin.

Every plumbing system must be vented to the open air and every fixture must have a water seal.

Every plumbing contractor must be licensed.

2.4.9   Heating

Codes are generally silent on heating capacities. In the case of multiple dwellings, there are regulations requiring minimum .interior temperatures during the cold seasons. Codes require that heating devices be safe and specify devices that must be installed. In the case of gas or oil heating, most codes require a licensed installer.

2.4.10   Electricity

The National Electrical Code is used throughout the country as a standard. It specifies the sizes of wiring, the size of services, safe fuse boxes and junction boxes, sizes of conduit, and all other matters relating to the safety of an electrical system. Improperly installed and undersized electrical systems are one of the primary causes of fire. The builder must be sure that the electrical contractor is licensed.

2.4.11   Energy

Many codes now require that insulation be installed in all the exposed surfaces of residential buildings and specify the maximum heat loss that will be allowed through these surfaces. Figure 2.12 is taken from a state building code. To use this chart, the builder must ascertain the average annual degree-days for the area. This should be obtained from the local building inspector who enforces the code. *The value of U is the Btu per hour per 1 square foot of exterior surface per degree Fahrenheit.* For instance, in an area of the country where the average degree-days amount to 4000, the U should not exceed 0.25. Chapter 13 will explain exactly how this can be accomplished.

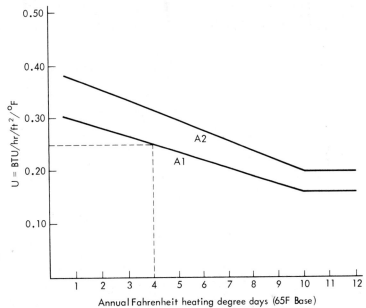

Figure 2.12. Maximum allowable U values for gross exterior wall assemblies.

As specified in Chapter 43 of the ASHRAE Handbook–Systems.
  Note 1: Line A1 inside graph denotes detached one and two family dwellings.
  Note 2: Line A2 inside graph denotes all other residential buildings not more than
      three (3) stories in height.

CHAPTER THREE

plans and specifications

The plans and the specifications for a dwelling must perform several functions. First, they must provide for a sturdy, weather-tight structure. Second, the plans should be so drawn as to provide a livable layout to suit the needs of the owner while not exceeding the owner's budget. This is a case where the architect, the builder, and the owner must exercise restraint. Third, the specifications should be written so as to ensure good workmanship and methods of construction and the use of good and available materials. The availability of materials is very important. Fourth, the plans must be drawn so as to comply with the building code and the size of the structure, and its placement on the building lot must conform to the zoning code.

3.1 PLANS TO SATISFY AN OWNER'S REQUIREMENTS

The builder, the architect, and the developer-builder must all be aware of the fact that almost every prospective owner would like more house than he or she can afford. The effort should therefore be made to plan a livable house within certain fixed parameters: those of cost and convenience.

There are certain rules for planning such a house, and if the house is convenient, much will be forgiven regarding the size. Some rules are:

The living room should be immediately adjacent to the front entrance, either directly or through an entrance hall (in a larger house).

If the living room is entered directly, the entry should be located so that the living room can be crossed in a straight path to the rear of the house (Figure 3.1).

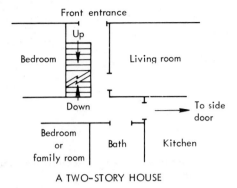

A TWO-STORY HOUSE

A ONE-STORY HOUSE

A ONE-STORY HOUSE

Figure 3.1. Ways of entering a house from front entrance without going through length of living room (in absence of center hall).

Every house has a front door and a rear door. The rear or side door, or even a second front door in some cases, is the working door. It should lead to the kitchen, the laundry room, or the pantry, and be near the basement stairs if possible. This is of utmost importance. It avoids muddy feet, delivery people, the carrying of groceries and supplies through the front door, and other inconveniences (Figure 3.2).

Bedrooms should open on a central hall, which should also lead to a bathroom.

The bathroom plays an important part in everyday living. It should therefore be as convenient and as large as possible. Figure 3.3 shows several alternative layouts.

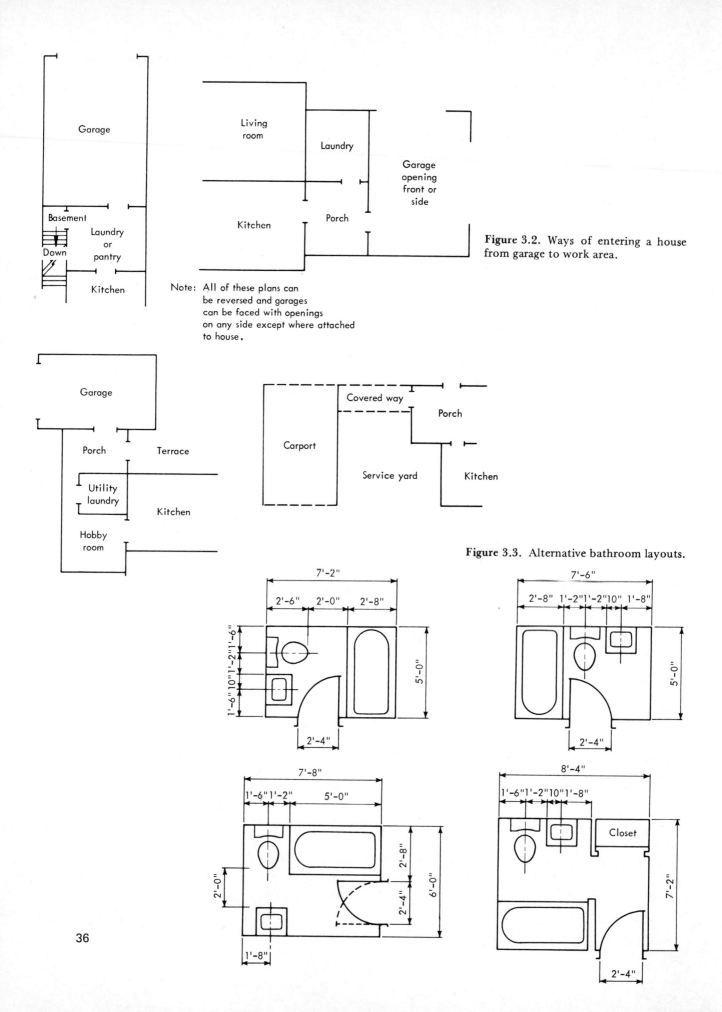

Garage

Basement

Down

Laundry
or
pantry

Kitchen

Living
room

Laundry

Garage
opening
front or
side

Kitchen

Porch

Figure 3.2. Ways of entering a house
from garage to work area.

Note: All of these plans can
be reversed and garages
can be faced with openings
on any side except where attached
to house.

Garage

Porch

Terrace

Utility
laundry

Kitchen

Hobby
room

Carport

Covered way

Porch

Service yard

Kitchen

Figure 3.3. Alternative bathroom layouts.

7'-2"

2'-6"   2'-0"   2'-8"

5'-0"

2'-4"

7'-6"

2'-8" 1'-2"1'-2"10" 1'-8"

5'-0"

2'-4"

7'-8"

1'-6"1'-2"   5'-0"

2'-0"

2'-8"

6'-0"

2'-4"

1'-8"

8'-4"

1'-6"1'-2"10"1'-8"

Closet

7'-2"

2'-4"

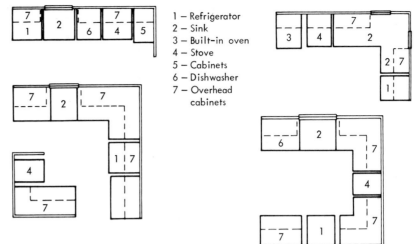

1 — Refrigerator
2 — Sink
3 — Built-in oven
4 — Stove
5 — Cabinets
6 — Dishwasher
7 — Overhead
      cabinets

Figure 3.4. Alternative kitchen layouts.

The kitchen work center is another important part of a house. As has been mentioned previously, a convenient layout and efficient working spaces can be planned for even a quite small, basic house. Figure 3.4 shows kitchen layouts in many possible shapes, and Figure 3.5 a well-planned L-shaped kitchen.

More than one door into a bathroom should be avoided.

Bedrooms should not be less than 10 ft in either dimension.

Almost all zoning ordinances require a garage or carport. This area should lead directly to the rear or working door so that the resident can enter the house from the car without being exposed to weather (Figure 3.2).

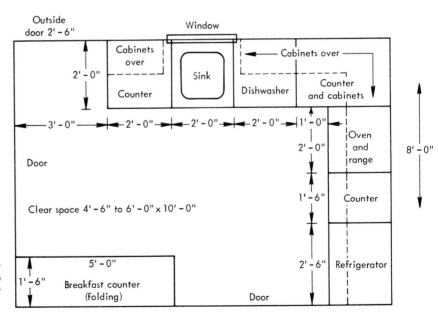

Figure 3.5. Detail plan of a medium-sized L-shaped kitchen to show how much can be done by good planning.

3.1.1   Plans for Varying Climates and Life-Styles

Although the room layout for convenient living is the same for every part of the country, the size and shape of the rooms can vary to accommodate different climates and life-styles. For instance, the use of a solarium or other open family space is desirable in a warm climate and a garage can be an open carport. Glass areas can be larger. The tendency is to live closer to the outdoors. Because there is no snow load, roofs need be only slightly pitched or flat. Construction can be lighter and heating plants can be minimal. Central air conditioning must be seriously considered.

In northern climates the living quarters are likely to have lower ceilings and smaller windows, to conserve heat. Roofs are usually more pitched. The emphasis is on indoor rather than outdoor living. There are regional styles of architecture from the so-called Cape Cod house to the thick-walled adobe house in the Southwest. The owner, architect, builder, or developer will find that the availability of materials has a great deal to do with the architectural style. In addition, the local prospective owner-builder is accustomed to local styles, and very few will want to build in widely different designs.

3.2   PLANS AND WORKING DRAWINGS

3.2.1   How to Obtain Them

Many owners and builders have thought that there is no reason to employ an architect or a designer to draw the plans for their houses, because they know what they want and can draw it themselves. This is true only to a limited extent. Certainly, every prospective home builder can decide how many rooms and what size rooms he or she wants. But how will this amateur decide on modular lumber sizes and how the structure is to be framed? How will he or she know that the strength and details of the framing meet local building requirements? This is also true for the plumbing, electricity, and heating. The owner may be able to get advice from a local carpenter, plumbing contractor, electrical contractor, and so on, and by becoming obligated to these people, will eventually produce a house plan. But that plan may or may not be accepted by the local building official.

There is also the important consideration of cost. A plan such as this would almost certainly lack precise detail, and lacking such detail, every contractor and subcontractor will, when estimating, allow for unknowns. This can also lead to on-the-job and after-the-job disputes about what the plans really meant. The services of an architect or designer (where the latter is allowed) are recommended. Such a professional should save more than the fee that is charged.

There is a way of obtaining plans and working drawings without going to the expense of employing an architect to design an individual house. Home-oriented magazines such as *American Home, House and Garden,* and *Better Homes and Gardens* publish catalogs of house plans that are available for purchase at reasonable fees. These plans are to an extent regionally oriented, and the reader should not try to build an adobe cottage in northern Minnesota! They also are drawn to comply with most building codes. Before they are used, the foundation and structural plans should be examined to be sure that they meet with the local code. The details of footing depths, foundations, exterior wall cladding, insulation, plumbing, and heating must be adapted to local weather. Such work can often be done personally by the builder or with the help of a local architect or designer (which is strongly recommended).

3.2.2   How to Adapt Purchased Plans

Purchased plans can be changed to meet the builder's or owners requirements or to meet the peculiar requirements of the building lot. For instance, the house plan shown here (Figure 3.6a) can be reversed. A stairway leading up to an attic and down to a basement can be installed by making the living room and the kitchen 3 ft longer. The shape of the bathroom can be changed. The garage can have room for two cars. There are other possibilities.

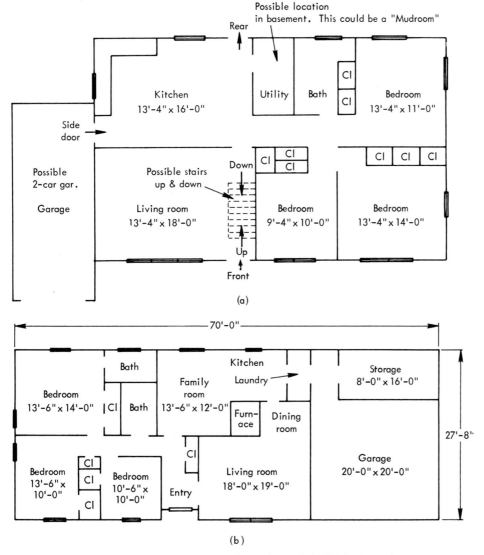

Figure 3.6. Sample plans that can be changed to suit individual requirements.

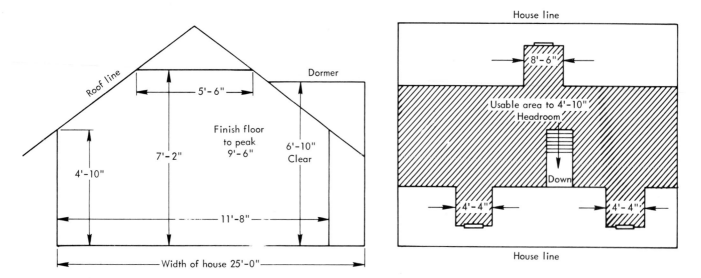

Figure 3.7. How to plan for an expansion attic.

The plan shown in Figure 3.6b can be reversed or turned sideways. The entry can be widened to make room for a stairway to a second floor and an attic, which will provide for future expansion. The two-car garage can be made smaller to provide for one car. Figure 3.7 shows how an expansion attic can be planned for very little extra cost.

3.2.3   How to Read a House Plan

The owner or builder, having now obtained a satisfactory set of plans whose room size and layout satisfies him or her, must now be able to interpret these plans. The directions must be followed carefully to obtain the structure as shown.

The first step in interpreting plans is to become familiar with the various architectural conventions and symbols that are used. Close attention must be paid to how the dimensions are shown. There are cross sections showing the exterior walls from foundations to roof and possibly of foundation details, fireplaces, and special structural framing. Different materials are shown by various crosshatchings or other marking when shown in plan (Figure 3.8).

There are also conventional ways of showing window and door openings, various kinds of windows and doors (in plan and in elevation), dimension lines, center lines, and others (Figure 3.9). Symbols for plumbing, heating, and electrical work are shown in Figures 3.10 to 3.12.

The method of dimensioning in frame buildings is shown in Figure 3.9. Dimensions are always taken from the structural stud. In window and door openings the dimension given is always to the center line of the opening. This enables the builder to compensate for any inequalities in the trim and permits changing the width of the opening right up to the moment when the frame is to be roughed in.

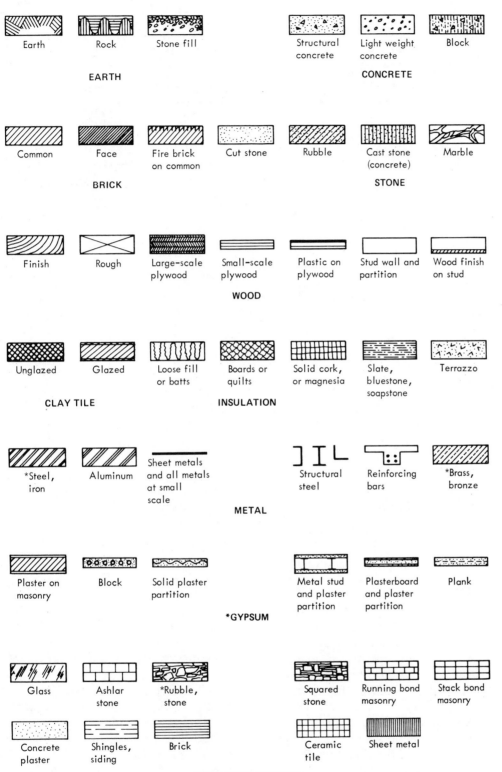

Figure 3.8. Architectural symbols.

Earth

Rock

Stone fill

EARTH

Structural concrete

Light weight concrete

Block

CONCRETE

Common

Face

Fire brick on common

Cut stone

Rubble

Cast stone (concrete)

Marble

BRICK

STONE

Finish

Rough

Large-scale plywood

Small-scale plywood

Plastic on plywood

Stud wall and partition

Wood finish on stud

WOOD

Unglazed

Glazed

Loose fill or batts

Boards or quilts

Solid cork, or magnesia

Slate, bluestone, soapstone

Terrazzo

CLAY TILE

INSULATION

*Steel, iron

Aluminum

Sheet metals and all metals at small scale

Structural steel

Reinforcing bars

*Brass, bronze

METAL

Plaster on masonry

Block

Solid plaster partition

Metal stud and plaster partition

Plasterboard and plaster partition

Plank

*GYPSUM

Glass

Ashlar stone

*Rubble, stone

Squared stone

Running bond masonry

Stack bond masonry

Concrete plaster

Shingles, siding

Brick

Ceramic tile

Sheet metal

SHOWN IN ELEVATION

*Shown in plan and section.

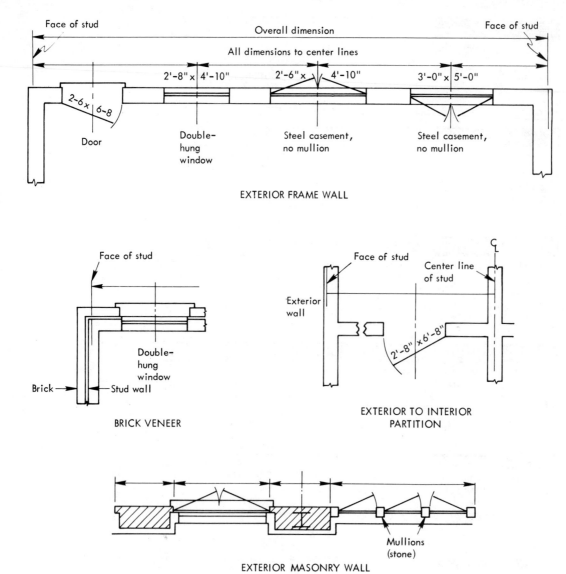

Figure 3.9. Typical dimensioning.

When the exterior wall is of masonry, the dimensions are taken from masonry corners or from the outer edges of stone mullions, as shown in the bottom drawing of Figure 3.9. The builder has some leeway in fitting the windows into the openings.

In dimensioning the elevations, the usual method is to dimension from finish floor to finish ceiling and to show the depth of the supporting joists. Dimensions are also given to the tops of the window and door openings. In calculating clear room height, the builder has to take into account the thickness of the finish flooring and finish ceiling material.

Most small-house plans that will be encountered by a builder will simply show the plan of the structure, a section through an exterior wall, the four elevations, a section through the foundation wall, and a framing detail.

There will be notes on the plans to show window and door sizes and the specifications will mention what kind of floors, wall finishes, ceiling finishes, and so on, are called for. Such plans will rarely show mechanical or electrical details and confine themselves to showing the location of electric outlets and ceiling or other fixtures; the location of heating plant and radiators or warm-air supplies and returns; and the location of plumbing fixtures and other plumbing outlets.

PLUMBING SYMBOLS			
Symbol	Plan	Initials	Item
————————	○	D.	Drainage line
— — — —	○	V.S.	Vent line
— — —	◎		Tile pipe
————————	○	C.W.	Cold water line
— — —	○	H.W.	Hot water line
— — —	○	H.W.R.	Hot water return
× × ×	○	G.	Gas pipe
●● ●● ●●	○	D.W.	Ice water supply
●●● ●●● ●●●	○	D.R.	Ice water return
╱ ╱ ╱	○	F.L.	Fire line
≻ ≻ ≻ ≻	⊕	I.W.	Indirect waste
— I — I —	⊕	I.S.	Industrial sewer
— ╲ — ╲ —	⊗	A.W.	Acid waste
— ·o· — ·o· —	Ⓐ	A	Air line
— ∞∞ — ∞∞ —	Ⓥ	V	Vacuum line
⊏ ⊏ ⊏ ⊏	Ⓡ	R	Refrigerator waste
●——▼—			Gate valves
↦——↦—			Check valves
⊏CO ⅄CO		CO.	Cleanout
☐ F.D.		F.D.	Floor drain
⊙ R.D.		R.D.	Roof drain
□ REF.		REF.	Refrigerator drain
⋈		S.D.	Shower drain
⊙ GT		G.T.	Grease trap
⊢ S.C.		S.C.	Sill cock
⊢ G.		G.	Gas outlet
⊢ VAC		VAC.	Vacuum outlet
⊢Ⓜ⊣		M	Meter
☐ˣ			Hydrant
▨ H.R.		HR	Hose rack
▨ H.R.		H.R.	Hose rack–built in
		L.	Leader
Ⓗᵂᵀ		H.W.T.	Hot water tank
Ⓦᴴ		W.H.	Water heater
Ⓦᴹ		W.M.	Washing machine
Ⓡᴮ		R.B.	Range boiler

Figure 3.10 (a). Standard plumbing symbols.

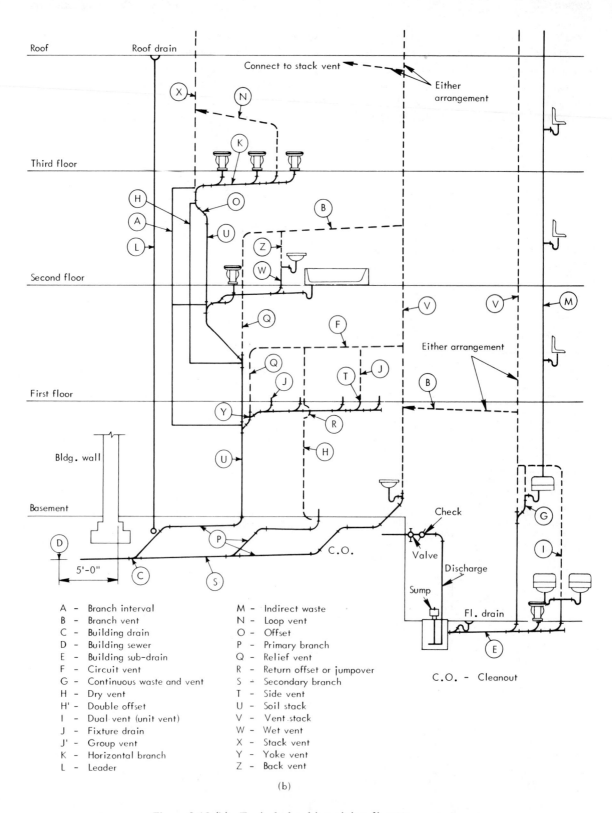

Roof — Roof drain

Third floor

Second floor

First floor

Bldg. wall

Basement

Connect to stack vent — Either arrangement

Either arrangement

Check
Valve
Discharge
Sump
Fl. drain

5'-0"

C.O.

A – Branch interval
B – Branch vent
C – Building drain
D – Building sewer
E – Building sub-drain
F – Circuit vent
G – Continuous waste and vent
H – Dry vent
H' – Double offset
I – Dual vent (unit vent)
J – Fixture drain
J' – Group vent
K – Horizontal branch
L – Leader

M – Indirect waste
N – Loop vent
O – Offset
P – Primary branch
Q – Relief vent
R – Return offset or jumpover
S – Secondary branch
T – Side vent
U – Soil stack
V – Vent stack
W – Wet vent
X – Stack vent
Y – Yoke vent
Z – Back vent

C.O. – Cleanout

(b)

Figure 3.10 (b). Typical plumbing piping diagram.

Figure 3.11.
Heating symbols.

Figure 3.12.
Electrical symbols.

The electrical work is always done by an electrical contractor, who should perform it in accordance with code requirements. The heating and plumbing sizes and appliances are usually designed by either the contractor or a professional who advises the contractor (see Chapter 17).

Where complete detail is not shown, the builder and the prospective owner should be very careful when making up or receiving an estimate to be sure that every item is covered.

The plans for a small house shown on pages 46 to 49 are an excellent example of what complete architectural plans should be. They should be studied very carefully to note how the architect has dimensioned the room sizes, door and window openings, floor heights, and structural sizes, and how materials are indicated. The plan should show graphically exactly what is wanted.

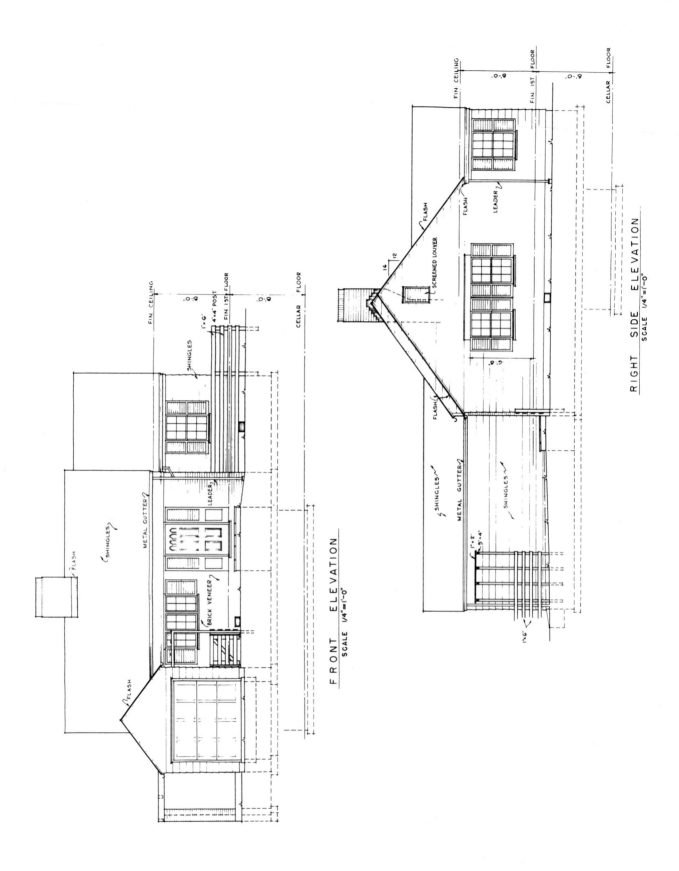

FRONT ELEVATION
SCALE 1/4"=1'-0"

RIGHT SIDE ELEVATION
SCALE 1/4"=1'-0"

46

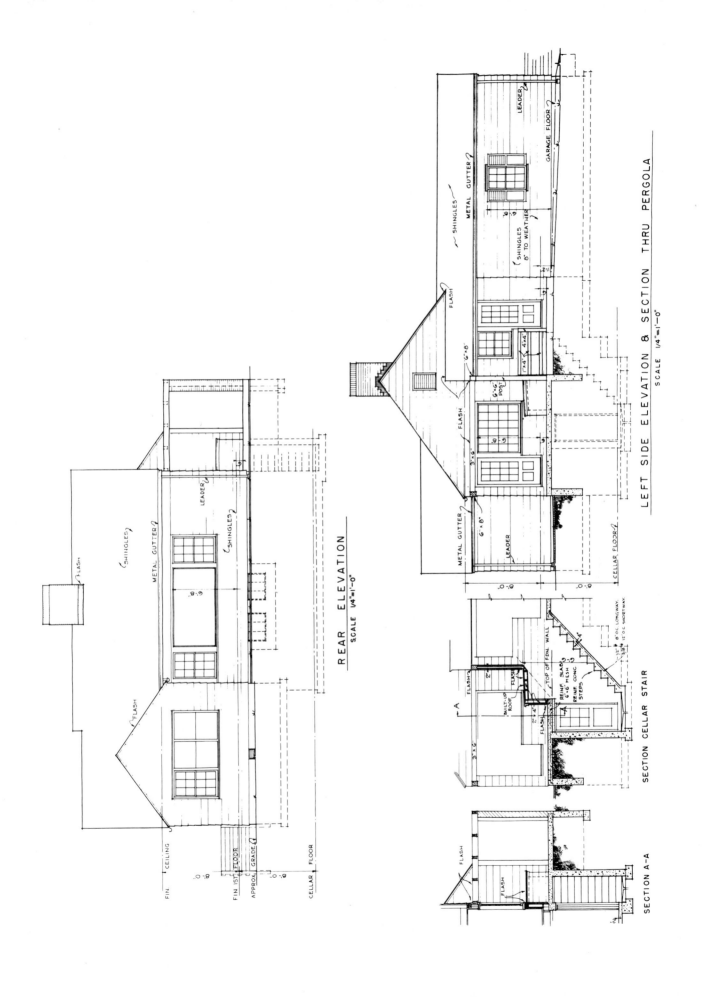

REAR ELEVATION
SCALE 1/4"=1'-0"

LEFT SIDE ELEVATION & SECTION THRU PERGOLA
SCALE 1/4"=1'-0"

SECTION CELLAR STAIR

SECTION A-A

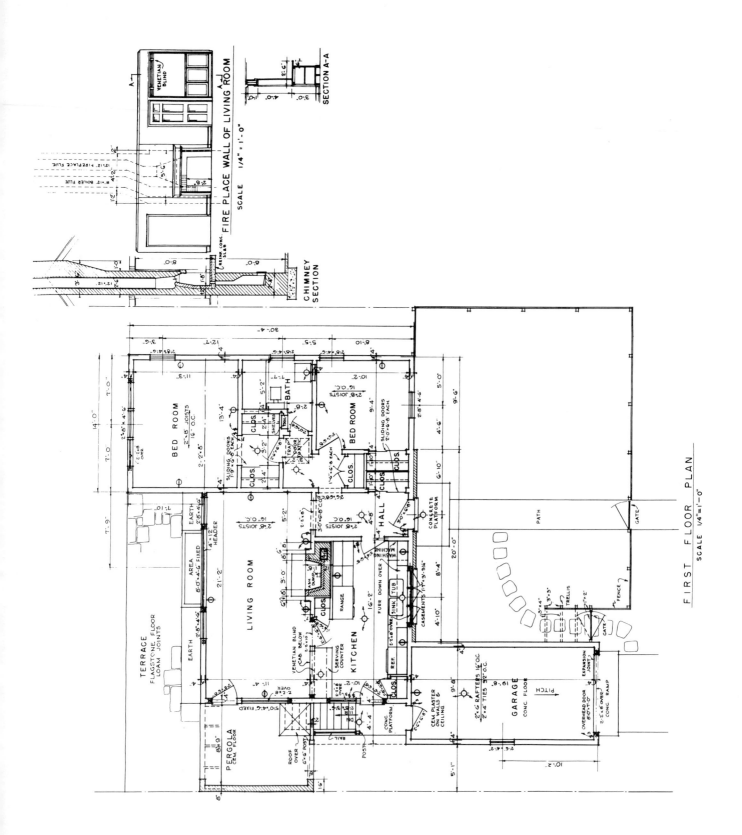

FIRE PLACE WALL OF LIVING ROOM
SCALE 1/4" = 1'-0"

SECTION A-A

CHIMNEY SECTION

FIRST FLOOR PLAN
SCALE 1/4"=1'-0"

48

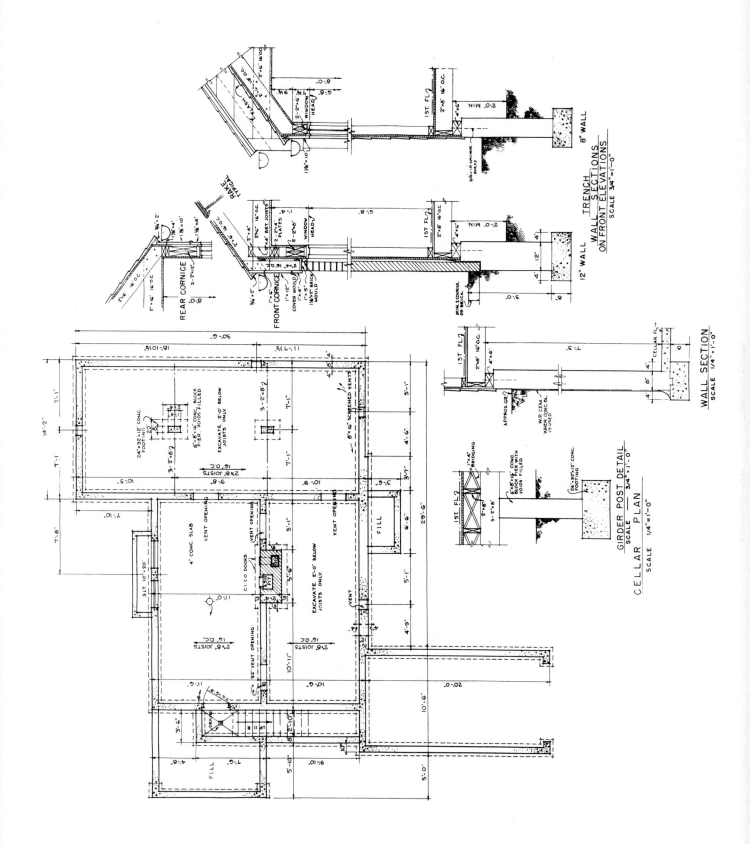

3.3 SPECIFICATIONS

The specifications for any proposed structure, together with the plans, form a complete whole, and neither portion is complete without the other. The plans show shapes and sizes and architectural and mechanical detail, but the builder also needs to know the material and quality of which the various members consist and how the parts of the structure are to be put together. The specifications also define the builder's obligations. Without a clear specification to complement the drawings, both the owner and builder have to guess at the type and quality of materials to be used, and this can lead to all sorts of disputes and unforeseen expenses.

3.3.1 Samples of Typical Residential Specifications

The specifications for the construction of a residential structure follow the order as formulated by CSI, The Construction Specification Institute, Inc., 1150 Seventeenth St. N.W., Washington, D.C. 20036. Their formats and basic specifications can be obtained by registered architects and engineers. Only the sections that apply to residential construction have been included below.

Division 1: General Requirements

The General Requirements define the obligations of the builder, the owner, and the architect. They will state that the plans and specifications are complementary and must be correlated with each other. It is the obligation of the owner to furnish a survey that exactly defines the property; to furnish any legal documents referring to easements, rights of way, and so on; and almost always in residential work to carry the fire insurance during the course of construction.

It is the obligation of the architect to interpret the drawings and to see to it that all work complies with the plans and specifications.

The builder is obliged to carefully study and evaluate the intent of the plans and specifications and to call the attention of the architect or owner to any discrepancies. The builder must also have examined the site and have stated that it is satisfactory to build on. In this connection, if the builder has any suspicions of ledge rock, water, unsatisfactory foundation conditions, or foreseen expensive land clearing, they should be called to the attention of the owner or architect before work is started. The builder is also obliged to carry proper insurance, for which the minimum requirements are workmen's compensation, employer's liability, comprehensive general liability, and comprehensive automobile liability.

Division 2: Sitework

Accurately stake out the foundations and determine the levels.

Clear the site as directed and strip and stockpile topsoil.

Excavate to undisturbed soil to the level below the frost line for footings and excavate generally as shown.

Excavate and backfill for all trades.

Establish subgrades for the basement slab and driveway.

Supply and distribute backfill where called for.

Rough-grade the site after completion to prepare for finish grading as shown.

Division 3: Concrete. Division 4: Masonry

Footings must be accurately laid out according to plan and must be laid on solid undisturbed frost-free earth. The quality of the concrete is specified by proportion of mixture or by strength. The size and shape of the footings is shown in the plans. Any reinforcement of the footings is specified by size and quality.

Foundation walls are shown on the plans. Their manner of being poured and reinforced is specified, as is the mixture and strength. In the case of block walls, the size of the blocks and the manner of laying them is specified. The kind of mortar is specified. The number and size of hold-down bolts is specified.

Drain lines around foundation footings are usually specified under Excavation, but the foundation contractor must be sure that they are installed.

Fireplaces, chimneys, and other masonry work are shown in the plans, but the kind and quality of brick and its manner of laying must be specified: for instance, "All brickwork to be set in running bond."

The manner of waterproofing and foundations is specified: for instance, "Apply ½-in. parging to all exterior concrete block foundations and apply two coats of asphaltic mopping."

The contractor must supply all foundation vents, chimney and fireplace appliances, cleanout doors, and an ash drop.

Division 6: Wood and Plastics

The kind of structural framing, the size of the members, and the method of joining them is shown in the plans. The specifications set forth the kind and quality of all rough and finish lumber, plywood, exterior siding or shingles, doors, windows, and trim—both interior and exterior.

The contracting firm should look carefully to see whether in addition to the framing and trim they are obligated to supply and install garage doors; furnish and install interior wallboard partitions and ceilings; furnish and install finish flooring; furnish and install screens and storm sash; and so on.

A sample of specified quality would read thus: "Roof decking to be size shown, 6-in. nominal; Potlatch lock deck, select Douglas fir, laid random length"; or "Siding: D grade or better resawn face western red cedar of section as shown."

Division 7: Thermal and Moisture Protection

When a roof is pitched, the contractor must first note which trade is doing the roof shingling. Wood shingling is done by the carpenter, but asphalt shingling can be done by a roofer or a carpenter.

Flat roofs are always done by a roofer because the specifications in such cases call for a three- or four-ply roof laid in hot pitch, which is roofers' work.

Weatherstripping of windows, doors, or other openings by a carpenter.

Gutters and leaders are specified as copper or aluminum of certain gauges. Many houses have wood gutters, which is carpenters' work.

The specifications will state where flashing is required and of what material and weight it should be. This is a case in which the plans may show flashings at certain places but the specifications do not mention them. The rule is that if anything is shown on the plan and not mentioned in the specifications, or if something is mentioned in the specifications but not shown on the plan it is the builder's obligation to do it.

The specifications may call for block-type insulation, such as foam glass or urethane under flat roofs (and sometimes under pitched roofs), and for wall and ceiling, floor, basement, and attic insulation.

Division 8: Doors and Windows

The doors and windows are specified by grade and thickness. Window specifications very often mention the preferred (or demanded) manufacturer. Many plans show schedules of windows and doors by size, location, grade, finish and other desired characteristics. If there are no schedules, then the elevation and plan will show sizes and possibly thicknesses and other essential information.

Glazing. The thickness and quality of the glass is specified. Normally, window sash comes ready glazed with A-quality double-thick glass.

The builder should look for any other glass to be furnished, such as mirrors or special doors.

Division 9: Finishes

Painting. A finish schedule is of great help to the painting trade. If there is no schedule, the specifications should describe the work by location and finish.

The preparation of all surfaces, prior to painting, should be mentioned first.

Exterior walls: The finish, whether paint or stain, should be specified by manufacturer, quality, and number of coats.

Exterior trim: Backpainting and quality of finish.

The specifications may mention all interior walls and ceilings to be finished with "two coats latex-based wall paint as manufactured by X Co."; kitchen and baths and interior trim "one coat primer and one coat semi-gloss enamel."

The builder must look for other work that is included, such as door finishes, special finishes, closets, hardwood floors, wood decks, pipe rails, any exterior metal work, and other things that usually require painting.

In most small projects the specifications will call for the painting contractor to tape and spackle all interior wallboard preparatory to its being
painted. This is an expensive and time-consuming job. In many large
towns there are special contractors who do only this work. This could
be investigated.

Tile. Ceramic tile is used most often in bathrooms, on walls, and sometimes on floors. It may also be used on kitchen walls. Some type of slate
or quarry tile may be specified for entry-hall floors. The plans will show
the location and size of the work.

A sample of quality follows: "Bathroom wall tile to be $4\frac{1}{4}$- by $4\frac{1}{4}$-in.
cushion-type tile as manufactured by Blank Co. Color to be Tan No.
123 matte glaze and to be set in adhesive as manufactured by Jones
Co. Contractor to furnish and set soap and grab at each tub."
The plans should show where the various types of tile are to be used.

Division 15: Mechanical

Very few plans for private residences show the details of the plumbing,
heating, air-conditioning, and electrical systems. Such details would show
the size and runs of water lines, valves, soil and vent lines, and so on, for
plumbing. They would show the location and size of heating ducts or heating lines. The electrical circuits would be shown with the location of the
main switchboard, panel boards, home runs, and so on.
Instead, usually the plans confine themselves to showing the location
of radiators or warm-air outlets, of plumbing fixtures and appliances, and
of electrical outlets and of appliances that use electricity. The matter of
size of water or soil lines, heating plant and radiators, ducts, and circuit
wiring is then left to the judgment and knowledge of the specific trade
subcontractor subject to code requirements. In such cases the specifications
bear the responsibility of defining the work to be done in a way to assure
a livable housing unit.

Plumbing

The specifications may say: "The system shall consist of soil, waste,
drain, vent, and hot- and cold-water supply lines. Valves are to be provided at all fixtures."
The specifications will state that the work conform to all local, state,
and federal codes.
The specifications may also say that all water lines should be so sized
as not to inhibit flow at one fixture when another is in use.
The material and quality of all piping and valves is mentioned; for instance, cast iron, copper, red brass, PVC, valves, as manufactured by X Co.
The manufacturer's name and catalog number of each fixture is mentioned.

The contractor should look for items such as roof drains, water heaters,
well pumps, and storage tanks, and (very important) whether there is a

street sewer or a septic tank and drain field. Connection to street sewers is plumbers' work, but a septic system may be installed by a special contractor. The specification should say who is responsible.

Heating and Air Conditioning

The specifications should mention the manufacturer, the type and catalog number, and the capacity of the central heating plant and the fuel to be used.

The specifications should mention the type of heating system, such as one-pipe or two-pipe forced hot water; one-pipe or two-pipe steam, forced warm air; and so on.

The specifications should mention the type of radiators or registers to be used.

The specifications should mention the requirements for heating, such as: "The heating system should have the capacity to heat the interior to a temperature of 70° Fahrenheit with an outside temperature of 0° F."

In the case of central air conditioning, the specifications should mention the required performance standards of the system, such as "an indoor temperature of 78°F and 50% relative humidity at an outside temperature of 95°F dry bulb and 75°F wet bulb.

Division 16: Electrical

A general phrase will say: "The work shall comply with all applicable codes and all wiring should be UL approved." It will also say: "The electrical contractor shall be responsible for all wiring in the house, including the connection of all appliances, including lighting fixtures, and heating and air-conditioning system, and shall furnish all necessary material to accomplish this."

The specifications will give the size of the service, such as 120/240-volt, 100-ampere, single phase. It will mention the manufacturers of the panel board, switch gear, outlets, plates, wall switches, and other wiring devices.

Such extras as door bells, light dimmers, telephone wiring, appliances that require extra amperage or voltage, and bathroom or other fans should be looked for in the plans and specifications.

CHAPTER FOUR

general principles of cost estimating

Cost estimating in construction is at best an inexact science. Exhaustive time studies of labor productivity show that a worker will perform a certain amount of work per hour or per day, and such figures can be taken as a general guide. But unlike assembly-line production of any kind, the construction industry has never reached this stage. It is true that some large developer-builders have produced assembly-line houses, but this is not true for the great majority of builders. The work is performed outdoors and is subject to the vagaries of weather. The material for construction normally cannot be stocked on the site very far in advance. There is usually not enough space, and in any case it would probably either be stolen or be spoiled by the weather.

An estimate is made up of several parts in addition to the cost of labor. There is the overhead cost of doing business, equipment cost, cost of material, insurance, and other on-site job costs. All of these costs can be figured fairly accurately. For the cost of labor, the builder must to some extent rely on labor productivity studies as well as on personal experiences with local labor and local construction customs.

Several of the following chapters go into some detail on estimating material quantities and labor costs, especially Chapter 8, Excavation and Foundations; Chapter 9, Structural Framing; and Chapter 10, Exterior Wall Cladding. These trades are the ones the normal builder is usually concerned with because he or she can deal directly with most of this work.

This book does not attempt to be a complete "how to" guide to estimating for all trades. There are a number of books that deal with such estimating. The titles listed below were selected somewhat at random from lists of books in print. The builder or owner can find additional titles from lists which are available in any public library.

Building Construction Estimating, 3rd ed., by George H. Cooper and Stanley Badzinski, Jr. New York: McGraw-Hill Book Company, 1971.

Estimating Building Construction: Quantity Surveying, by William Hornung. Englewood Cliffs, N.J.: Prentice-Hall, Inc., 1970.

Fundamentals of Construction Estimating and Cost Accounting, by Keith Collier. Englewood Cliffs, N.J.: Prentice-Hall, Inc., 1974.

National Construction Estimator, by Gary Moselle. Solana Beach, Calif.: Craftsman Book Co., 1977.

The correct estimating of the cost of the construction work to be done is one of the most important parts of the construction process. The builder who bids on a construction job and who subsequently obtains the contract to perform the work can stand or fall on the correctness of the estimate. The owner who wishes to be directly in charge of part of the work must also know what the work will cost. The construction of even a small private residence involves many trades and materials, and the cost of many of these may come as an unpleasant surprise and catch the owner financially unprepared.

4.1   DETERMINING THE INTENT OF THE PLANS AND SPECIFICATIONS

Before the start of any construction, it is strongly urged that the builder, whether the homeowner or a professional, determine the intent of the plans and specifications. The questions asked should be somewhat as follows:

Does the architecture of the house call for any special framing?

Is the structure to be built to just comply with the code (which is minimal), or will it be better? (A builder or architect will know the answer to this.)

What is the general topography of the site? Does it look as though there will be excavating problems, such as ledge rock, boulders, water, or unstable foundation conditions?

What quality of house does the owner want and is he or she able to pay for? (Some brand names are advertised as being the very best, but other products may be just as good and cost less.) Experience or informed opinion will ferret these out. Inquiries at good material dealers will provide such information. If an architect is used, he or she should be carefully questioned as to the quality of material being specified.

4.1.1   The Various Trades and the Work Involved

The first step in cost estimating is to determine what trades are involved and the extent of the work to be done by each. The building firm must now decide what work it can perform, perhaps with the aid of additional help, and what part it must subcontract. It will be of help to refer to the outline specifications shown in Chapter 3.

General Conditions

This item of the cost of construction is called General Conditions because it is made up of many pieces of work that must be done by the contractor, the owner, or a subcontractor, but which do not specifically fall into any trade. Some but not all of these items are:

Rubbish removal.
Cutting and patching to repair work damaged by any trade.
Temporary weather protection during construction.
Temporary water supply.
Temporary sanitary facilities.
Winter protection during construction.
Special scaffolding.

The combined cost of these items can come to a sizable sum and the estimate must take account of this.

Miscellaneous

Like General Conditions, this item is a catch-all. In this case the work to be done is usually performed by a specific trade, but in the usual residential construction each item is too small to justify employing a specific sub-contractor. The following items should be taken account of and the owner or builder must either assign them or be prepared to install them personally:

Miscellaneous metals, such as steel lintels and lally columns.
Structural metals, such as columns, girders, and channels.
Weatherstripping and caulking.
Weather insulation.
Sound insulation.
Drywall partitions.

Sitework, Excavation, and Foundations

Clearing the site may involve trucking of heavy timber and brush. Excavation will involve power machinery such as power shovels and heavy trucks. Foundations will involve concrete and formwork or the laying of concrete block. It may also involve waterproofing of below-grade walls and drainage. The builder has the choice of doing any part of this by hiring day labor or by itemizing a list of work to be done and obtaining a bid for the entire work from a subcontractor. In the latter case, more than one bid should be obtained.

Masonry

Masonry involves such items as chimneys and fireplaces, front and rear steps, patios, brick or stone veneer walls, and the foundation walls. The contractor who bids on the foundation walls can also be asked to bid on all of the masonry work. Here again much of this work can be done by day labor except chimneys and flues, which are strictly inspected and should be built by expert masons.

Carpentry and Millwork

This work is always done by the professional builder. The owner who wishes to do a good part of this work is well advised to employ a carpenter contractor to at least build the structural framing and enough of the other work to make the structure reasonably weather-tight. This might involve the sheathing and tar-papering of the roof and the sheathing of the walls. This is not only important for the weatherproofing but also adds necessary structural strength to the house. A bare structural frame should not be left as is for any length of time. Interior partitions, except bearing partitions, can be installed by the owner. Wallboard installation and taping requires a certain skill.

Roofing and Sheet Metal

This work involves the installation of the final weatherproof roof covering; the leaders, gutters, flashings; and other material to shield the interior from weather. If the roofing is of asphalt or wood shingles, it can be installed by the carpenter contractor or the owner. If it is a flat roof or a slate roof, it is wise to employ a professional who is familiar with this type of work and who owns the proper equipment.

Ceramic Tile

If the owner or builder requires ceramic tile in the bathroom or kitchen, it is recommended that it be done by a professional tile setter. There are, of course, many substitutes for ceramic tile, such as sheet plastics, which can be installed by a carpenter or even by the owner.

Painting and Finishing

This work can be done by the owner, but he or she is advised to carefully investigate the quality and type of the paints to be used. Paint manufacturers publish brochures which tell which paints they recommend for various surfaces. There are water-based latex paints, alkyd resins, and others for both interior and exterior use. Special care is recommended in choosing the paint for exterior walls and trim. It is best to get advice from a professional who is not trying to sell anything. The taping of interior wallboard prior to painting is a job requiring special skill. Poor taping will show through the painted finish and is very expensive to repair. Unless an almost invisible joint can be obtained, it is recommended that the work be done by a professional.

Plumbing, Heating, Air Conditioning, and Electricity

All of these trades require the use of licensed contractors and are strictly inspected by building inspectors. The reason is that all these trades have to do with health and safety. The inexperienced worker can easily omit a back-water valve, which could result in the mixing of wastewater with the drinking supply. Undersized electric wiring or bad connections can and have caused many fires. A heating plant can be undersized, the burner may not be properly installed, or the warm-air ducts may leak or make whistling noises. It is strongly recommended that the basic plumbing, heating, and electrical work be done by licensed professionals, as called for by all codes. The *knowledgeable* amateur can make some minor final connections.

4.2 REQUIRED INFORMATION BEFORE ESTIMATING

The first step in the pricing of the work is to break down each item of work by material and labor and then in the summary to add a percentage for overhead, profit, insurance, workmen's compensation, and other items chargeable to the work. The building firm that goes into competitive bidding or must prepare an estimate of the cost of the work before it is awarded a contract must determine the cost of doing business and what profit it wishes to make over and above expenses.

4.2.1 The Builder's Overhead

Overhead represents the builder's cost of doing business—costs that are not chargeable to any particular job. The following is a list of overhead items. The individual builder can add or subtract from this list.

Office rent—if the builder works from home, charge part of the house expense.

Light, heat, and power—if not included in the rent.

Telephone—answering service; job telephone to be charged to individual job.

Stationery and printing.

Office supplies.

Insurance—not chargeable to individual jobs.

Interest on borrowed funds.

Bonding.

Office salaries—add Social Security and other benefits.

Field salaries—not chargeable to particular jobs.

Accounting fees.

Legal fees.

Travel expenses—include mileage charge on automobile.

Depreciation of office furniture and equipment.

Subscriptions to publications; charitable contributions.

Association dues.

Rental or cost of maintaining storage yard, including taxes, mortgage interest, and interest on investment.

Depreciation of working equipment.

4.2.2   The Owner-Builder's Overhead

It should be noted here that the owner who proposes to perform any part of the work also has overhead expense. There are telephone calls, trips for material, fire insurance during the course of construction, legal fees, cost of any temporary supervision, interest on any borrowed funds, possible payment for equipment, and other indirect expense. This cost should be added to the total cost of the house.

4.2.3   How to Assemble a Bid

There are many available choices of materials to be made and many choices regarding such items as foundation walls, grading, exterior walls and roofs, interior finishes, and plumbing, heating, and electrical service. Many of these choices will be shown and explained in succeeding chapters. *It is at this point after the choices have been made in each trade category, that there will be a guide to estimating the cost of the work under this category.* As they go through the items, the builder and owner should decide what part of each trade they wish to do themselves and list everything else for bid.

Figure 4.1 shows a sample bid list which can be followed in putting an estimate together. Note that this bid is only a summary. The estimates for each trade that go to make up this summary will be received as lump sums from the various subcontractors that are asked to bid on part or the whole of that particular work. Where the owner and builder decide what part of the work they will do themselves, they have to make an estimate of this, which will include labor, material, overhead, rental of equipment, and use of tools. For instance, for foundation work there is the labor cost and the material for forms (in the case of concrete walls), the cost of the concrete or the blocks and mortar for the foundation wall, and the cost of hold-down bolts (where required), drainage tile, and waterproofing. For framing there must be a lumber list—all of which will be shown in succeeding chapters.

4.2.4   How to Deal with Subcontractors

The professional building firm in its dealings with subcontractors has the advantage of the possibility of repeat business. The subcontractors hope that if their price is reasonable and their workmanship satisfactory, they will be considered for further work. This is not so for the "one-shot" owner who wants to do part of the work himself or herself. He or she will not get the trade discounts for materials that the professionals get or certainly not all of them, and he or she may not get the same service and prices from the subcontractors.

To answer this ever-present problem, the owner-builder or the professional builder must do certain things:

Be sure that everything subcontractors must do is put in writing, including the quality of the materials.

Be sure that every item necessary to make a habitable house is in someone's contract or be prepared to do it themselves.

Obtain more than one bid.

SUMMARY OF ESTIMATE

Building _ _ _ _ _ _ _ _ _ _ _ _ _ _ _ _ _ _   Address _

Owner _ _ _ _ _ _ _ _ _ _ _ _ _ _ _ _ _ _ _   Address _

Architect _ _ _ _ _ _ _ _ _ _ _ _ _ _ _ _ _ _   Address _

Classification	Estimated Labor Cost	Estimated Material Cost	Subcontractors' Bids	Estimated Total
General conditions, overhead				
Permits, insurance, taxes				
Supervisor, foremen, watchmen				
Plant, tools, equipment				
Site clearing				
Excavating, backfilling				
Grading, rough and finish				
Foundations				
Waterproofing, tile, gravel				
Walks and driveway				
Brick, tile, concrete, masonry				
Rough carpentry				
Insulation, thermal, sound				
Finish carpentry, millwork				
Floors, laying, finishing				
Stairs				
Rough hardware				
Finish hardware				
Weatherstripping				
Caulking				
Drywall, lath, plaster				
Sheet metal, flashing				
Gutters and downspouts				
Roofing, wood, asphalt				
Structural iron				
Miscellaneous iron				
Tile and marble				
Glass and glazing				
Painting, exterior				
Painting, decorating interior				
Plumbing				
Heating, air conditioning				
Electrical				
Lighting fixtures				
Screen, storm doors				

ESTIMATED TOTAL COST _ _ _ _ _ _ _ _ _ _ _ _ _ _ _

OVERHEAD _ _ _ _ _ _ _ _ _ _ _ _ _ _

PROFIT _ _ _ _ _ _ _ _ _ _ _ _ _ _

BID PRICE _ _ _ _ _ _ _ _ _ _ _ _ _ _

Figure 4.1. Summary of estimate.

Before letting the contract, the builder must go over the job with subcontractors to make sure that they are in accord.

4.2.5   How to Assign and Price Miscellaneous Work

There have been numerous references to "Miscellaneous Work" and there has been a partial list of such work in Chapter 3 and in Section 4.1. The simplest way to take care of such items is to assign them to subcontractors while asking them to quote. For instance, the interior carpentry subcontractor can be asked to quote a price for erecting and taping interior wallboarding ready for painting. The foundation subcontractor can be asked to quote on parging (coating with cement mortar) the exterior of the basement walls, to trowel-coat it with a mastic compound, and to quote on foundation drainage. The carpentry subcontractor can be asked to quote on thermal or sound insulation. Roofing may be installed by a carpenter or a roofer. But leaders, gutters, and flashings are roofers' work and must be so assigned and priced. There are many other items of this kind, and many will be shown as the process of construction is explained in detail.

Nothing prevents the builder or the owner-builder from doing any of these things themselves if they feel they have the skills and time. In such cases, they must make a material list and obtain prices on specific items.

CHAPTER FIVE

the business of construction

5.1 BUSINESS PRACTICE IN CONSTRUCTION

Construction is a complicated business, and builders who wish to stay in business must be familiar with its basic business principles. Even the owner-builder of a single house should be familiar with these principles. The actual physical construction of a building is a very important part of the business, but it is only part of the whole. Following are summaries of the essential areas of knowledge with which a builder should be familiar.

5.2 LEGAL ASPECTS

The builder should be familiar with the fundamentals of real estate and construction law.

5.2.1 Contracts

A contract is an agreement enforceable by law. To be enforceable, the contract must consist of an *offer and an acceptance.* It is not enough for contractors to assume that they have been awarded a job because of what the owner has said, nor are they legally guaranteed the job even if the owner puts it in writing but they have not accepted it. The contracting firm may order materials and prepare to perform the work based on the owner's instructions, but if the firm has not signified its acceptance, the owner can still change his or her mind.

The function of a contract is to spell out the terms and conditions of an agreement between two or more parties. For instance, the owner wishes to sell and the purchaser wishes to buy a piece of land or the owner wishes to build and the builder agrees to perform this work. There are contracts between owner and architect, owner and builder, builder and subcontractors. The contract must name and identify the parties to it, the date of the contract, the work to be performed or the property or material that is involved, who will pay how much to whom and how, and when and on what terms payment is to be made. There are printed forms of contracts between owner and architect, owner and builder, builder and subcontractor, and others available. The American Institute of Architects, 1735 New York Ave. N.E., Washington, D.C 20006, has such documents available.

The owner, the builder, the owner-builder, and the subcontractor are all strongly advised to obtain a signed contract or its equivalent *in writing* before performing any service or obligating themselves to pay any money.

5.2.2   Construction Liens

Every state has a lien law which protects the rights of a contractor, a subcontractor, a material supplier, an architect, a worker, or anyone who has by his or her work or material added to the value of a property, to be paid for such labor and material. It is important to any contractor or worker that the person who authorizes the work to be done is the owner or an authorized agent. All states allow such a lien to take precedence over a mortgage taken after the work has been done, and many states allow a lien to take precedence over all mortgages.

Each state prescribes how soon a lien must be filed, where it must be filed, and in what form. The builder or worker who wishes to file a lien should seek legal advice. Lending institutions that make construction payouts ask the builder to sign a waiver of liens as a condition of being paid. The builder or owner-builder must in turn protect himself or herself. When he or she, in turn, makes a payment to any worker, subcontractor, or material dealer, he or she must be sure that they also waive their rights to a lien. Very often bills or checks marked "Paid in full" or payroll records are sufficient. The lien laws are complicated and the layperson should not file one until all other means of collecting payment are exhausted. The one thing to remember is that there is a time limit for filing a lien.

5.2.3   Local and General Business Laws

The purely local laws that the builder should be familiar with are usually the building and zoning codes, any licensing that is required for certain trades, and rules regarding storage of materials, sanitary facilities (both temporary and permanent), driveways, and connection to utilities.

Business laws are usually state laws. General business laws refers to the mode in which the contractor does business. He or she may be a sole proprietorship, whereby he or she is solely responsible for all debts but keeps all the profits. The sole proprietor can also be sued for debt, and anything he or she owns can be attached. The builder may also incorporate and file for a certificate of incorporation. A third choice is a partnership. Each form of doing business carries its own responsibilities, and legal advice is recommended before the builder decides how to do business. Labor laws, tax laws, and insurance laws make up part of business law.

5.2.4 Labor Law

Most states have their own labor laws. These refer to such matters as the employment of minors, the conditions of employment, job safety, and hours of employment. Federal labor law refers to such matters as discrimination against minorities, minimum wages, and the Occupational Safety and Health Act (OSHA). The latter law is becoming a significant factor in the construction business. OSHA inspectors of job safety are becoming more and more prominent. The builder who follows commonsense job safety rules should have no trouble with state inspectors or OSHA.

5.2.5 Tax Laws

The best way to write about the numerous and pervasive tax laws is to list them and explain their intent. The first and very important step in dealing with all tax laws is for the entrepreneur to keep accurate business records. (Record keeping is discussed in Section 5.4.)

Federal Income Tax

This law applies not only to the builder's own profit but also to the wages of his employees. He must file quarterly wage withholding statements (W-2's) and pay this money into a designated Federal Depository at stated intervals.

Social Security Law

The employer must also withhold money for Social Security and, after making his or her own contribution as set by the law, must pay such money into a Federal Depository on a quarterly or monthly basis. The Social Security law contains certain limits as to hours and costs of labor that must be reported. The self-employed builder who may not hire any help must also report and pay under the "self-employment tax" rule.

State and Municipal Income Taxes

A builder must be familiar with the requirements of the state and the local governments for their various income and general employment taxes.

Sales Taxes and Personal Property Taxes

States and municipalities levy sales taxes on the materials that the builder has to purchase. Many states also levy taxes on the contractor's personal property, such as large power equipment, automobile, and trucks.

Federal Unemployment Tax

This tax is a set percentage of the total payroll and applies to any employer whose payroll exceeds $1500 in any calendar quarter or any employer who has someone on the payroll for any part of a day for 20 calendar weeks.

5.3 INSURANCE

The builder must carry certain specified kinds of insurance coverage and should consider others.

Workmen's Compensation

By law, all employers must carry workmen's compensation insurance. Some states administer this insurance by public authority. In other states the insurance is carried by private insurance carriers. The law is very strict in every state regarding the filing of accident reports.

Fire Insurance

Fire insurance is usually carried by the owner and protects the owner and the builder against loss by fire and usually other risks, such as lightning, windstorms, explosion, and so on. The amount to be carried can increase as the work progresses or the cost of the entire completed work can be carried. The latter provides excellent discounts. The alteration builder must be sure that the owner carries a fire insurance policy that covers all material and partly completed labor.

Public Liability and Property Damage

The builder must carry this insurance to protect himself or herself and the owner against claims for personal injury or property damage that may be made by the general public.

Motor Vehicle Insurance

This protects the builder against claims for bodily injury or property damage caused by any motor vehicle used in performing the work.

The owner would be wise to assure himself or herself that the builder carries the appropriate insurance. Very often, suits for damages name the owner as well as the builder. In turn, the builder should be sure that all subcontractors carry such insurance. If they do not, the builder is liable.

There are other types of insurance coverages in addition to the ones mentioned here, but the builder or owner can consult an insurance broker to consider their cost against the risk involved.

5.4 RECORD KEEPING

5.4.1 Why It Is Necessary

It is generally agreed that the quickest way to lose a business is through the failure of the contractor to keep adequate records on a current basis.

As a matter of fact, the contractor is forced to keep financial records to satisfy the various governmental authorities, who may want to examine such records. A glance at the *Federal Tax Guide* reveals some highlights that are worth repeating.

You must keep records to determine your tax liability. . . . your records must be permanent, accurate, complete and must clearly establish income, deductions, credits, employee information, etc.

You must elect an accounting method . . . for income and expenses of the business.

If you have one or more employees you may be required to withhold Federal Income Tax from their wages. You may also be subject to the Federal Insurance Contribution Act (FICA) [which is Social Security] and the Federal Unemployment Tax Act (FUTA) [unemployment insurance].

These three quotes should be sufficient to show why a contractor must keep books. But there are other reasons, such as sales tax records, state and local income or real estate taxes, and so on.

5.4.2 How to Keep Records

This chapter does not presume to discuss general bookkeeping or accounting practice. However, it will show the contractor what records must be kept so that a person with a basic knowledge of bookkeeping can keep the books in order on a current basis and so that an accountant can go over the records and books at periodic intervals.

Following is a general outline of what an accountant requires to prepare a balance sheet, a statement of net worth (which is required by all lenders), and a profit-and-loss statement for IRS and others.

A payroll record, which conforms with all laws having jurisdiction. The record should include time sheets and hourly wages paid.

The inventory ledger, which records all supplies and materials on hand and shows the inflow and outgo of material.

The subcontractors' ledger, which shows all contracts made for the particular job, their cost, what has been billed, and what has been paid.

An equipment ledger, which shows all equipment on hand, when purchased, and how much it cost. Periodic inventories must be made to keep this record current. This record is also used for depreciation and insurance purposes.

The contract ledger, which records the various contracts entered into and their present status, including amounts paid out and amounts received on each contract.

The general ledger, which combines all the above and acts as a control on them. It is from this ledger that the accountant prepares the balance sheet and all the other records that show the "state of health" of the business.

5.5 FINANCIAL

The financial end of a business would seem to be so important that it should be a primary concern of any builder. But many small builders are so busy in the actual construction that they forget finances—and that way lies disaster.

Payrolls must be met; material must be paid for; subcontractors must be paid; insurance premiums must be paid on time; the list is long and with almost no exception charges must be paid within certain fixed times. Where is the money coming from?

5.5.1   Assets

There is, first, the builder's or owner's capital. The builder should first list his or her own assets, such as:

Assets (Liquid)

Cash in bank (savings)
Cash in bank (checking)
Stocks and bonds
Cash values of insurance policies

Financial interest in a property or business *that is readily convertible to cash.*

Assets (Fixed)

Automobile
Truck
Tools and equipment

The builder should then list his or her personal expenses, which represent an outflow of capital until he or she starts to have income.

Expenses (Personal)

Household expenses
Mortgage payments
Taxes
Electricity
Fuel
Telephone
Insurance
Car expense (gas and oil)
Miscellaneous expense (clothing, laundry, etc.)
Medical expense (possible)

5.5.2   Payment Schedules

If the builder is currently engaged in construction, he or she should also list as assets the money that is owed for work that has been completed, which includes construction payments. A typical construction payout schedule of a financial institution follows:

Payment 1

At the completion of the excavation and foundations and the start of framing.

Payment 2

At the completion of the framing and the start of rough plumbing and heating.

Payment 3

When the structure is enclosed and all mechanical and electrical roughing is completed.

Payment 4

When interior partitions, floors, and ceilings are completed ready for finishing and interior and exterior trim and cabinet work are under way.

Payment 5

At full completion, when a certificate of occupancy is issued.

The builder should be aware of this payout schedule and should have a clear understanding with the owner and the institution about when and to whom these payments are to be made. It is best to make it a part of the contract. The builder should apply for payment several days before he or she is ready and try to make a fixed appointment with the lending institution inspector to make sure that the money is received when it is due.

5.5.3   Liabilities

The next step in the financial picture is the preparation of a list of liabilities. To whom is money owed, how much, and when must it be paid? The builder must make a list of these items and *must* keep it current. The list includes:

Payroll.
Material bills.
Equipment rental.
Insurance.
Taxes.
Interest on loans.
Overhead.
The builder's own pay and personal expenses.

These items must be paid for from the assets, the construction payouts, and payments for miscellaneous work done. The money coming in and the money going out is called the "cash flow," and the builder must make every effort to keep this in balance so that there will always be some reserve working capital.

5.5.4 Types of Loans and Sources of Funds

There will be times in any builder's experience when because of delayed payments or delays in work that cannot be controlled, he or she is obligated to pay out sums that would seriously deplete or even wipe out working capital. In such a case the builder should be prepared to borrow money. There is no better way to build up a credit rating than to borrow money and repay it on time.

To borrow money, the contractor must be able to furnish complete personal and business information. Such information would include a financial statement or a balance sheet (usually prepared by an accountant), a profit-and-loss statement, and the contractor's record in paying bills. Even if this is the first time the contractor has asked for a loan, he or she will certainly have a credit rating on installment payments, mortgage payments, credit-card payments, personal loan payments, and so on. The lender will also want to know about the work in progress, the work that has been completed, and the prospects for future work. The lender may also want to see a projected financial statement for the period to be covered by the loan.

The Collateral Loan

It is possible that the lender, especially in the case of a new small business, may require collateral. This means that the loan is secured not by the business in hand and the contractor's past record of meeting obligations, but by the cash value of an insurance policy, stocks and bonds, improved real estate (e.g., the borrower's home), and savings accounts. The contractor should resist such a loan. Illness or death or any unforeseen catastrophe can cause the loss of a home or the loss of long-time savings.

The Personal or Character Loan

This is an unsecured loan which is based on the lender's appraisal of the contractor's ability to pay it back when called for. Such loans to a small business will usually run from 30 to 90 days before they come due. Banks acquainted with the contractor through having had his or her accounts or perhaps through paid-off automobile loans or home-improvement loans will usually give a sympathetic hearing to a request for a short-term loan for the purpose of providing cash flow for a going business. It is well for the prospective borrower to confine his or her accounts and personal borrowings to a single bank in order to build a credit and character record. The contractor must also remember that he or she is personally responsible for the repayment of such a loan.

The Installment Loan

This is somewhat the same as an installment loan on a car. This loan is discounted in advance; that is, the lender takes all the interest for the entire term of the loan out of the face value of the loan before the borrower gets the money. The contractor would then pay it back in monthly installments.

Very often, if the contractor has been faithful in making payments, when the loan is paid down he or she may be able to get it refinanced for an additional term and at a more favorable rate. This is almost the same as a line-of-credit loan.

Sources of Funds

Local commercial banks or building and loan institutions are the best sources for such funds. If the builder is well established, he or she can usually obtain a personal loan based on a record of financial reliability. The builder may even be able to get a "line of credit," or revolving loan, by which he or she can borrow up to a fixed amount and repay part of it, then borrow again—as long as the total loan never exceeds the maximum. There are also collateral loans which require the builder to pledge some of his or her capital assets. It is a good idea for a builder to make occasional loans and to repay them *on time.* In this way a good credit rating can be built.

There is also the federally funded Small Business Administration, which will guarantee loans to small business managers who cannot otherwise obtain credit. This agency has offices in every large city.

CHAPTER SIX

preparing for construction

6.1 BEFORE STARTING

Before starting construction the builder must obtain certain permits and approvals. Failure to obtain these at the very start can cause costly delays or even a complete work stoppage.

Obtain a building permit to make sure that the structure meets with building and zoning code requirements.

Obtain necessary insurance (fire, liability, workmen's compensation).

Arrange for job facilities, such as sanitary, telephone, and so on.

Have an electrical contractor obtain a power connection from the local utility.

Have a plumber obtain permits for water and sewer connections.

Obtain approval of a septic sanitary system if no sewer is available.

Clear the site and remove obstacles to construction.

6.2 SITE PLANNING

Site planning is the first step in the actual construction process and it is not an exaggeration to say that a great deal of the success of the project depends on good site planning. It is not a complicated technical process. It simply requires knowledge of where the house is to be located, what materials and what trades are to be used in its construction, and where all these trades are to work and their materials to be placed so that they will not be in the other's way.

6.3 CLEARING THE SITE; ZONING

Before clearing the site, the location of the house should be determined. This must be done within the zoning envelope, because zoning ordinances must be complied with even on large plots. On smaller plots it can be critical. Figure 6.1 shows a theoretical building lot of 10,000 ft². It is in a zone that allows a minimum building lot of ⅕ acre or 8712 ft². The diagram shows the minimum allowed side yard and the front and rear yards. The zoning envelope within which the house can be placed is 40 ft wide by 116.6 ft deep.

If the house is to be set with its front parallel to the street, it cannot exceed 40 ft in length, including the garage. It can, however, be set with the front at a right angle to the street. How and where the house is placed can depend on the slope of the land, the presence of ledge rock or large boulders, the presence of low-lying spots, sun direction, or other factors. The presence of ledge rock will cause expensive foundations and can even cause difficulty in digging ditches for utilities. Wet spots can sometimes be eliminated by proper drainage. It is best to avoid being under large trees.

When the best area for the location has been found, the next step is to clear the land. This will require labor and possibly equipment.

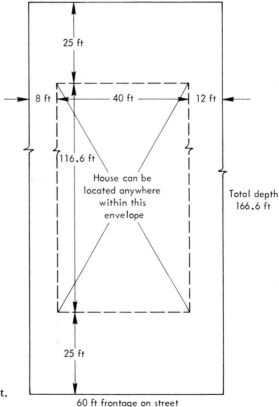

Figure 6.1. Location of a house on a small plot.

Building lot is 10,000 ft² with 60-ft frontage. Depth is 166.6 ft. Zoning envelope is 40 ft by 116.6 ft.

73

6.3.1   Estimating the Cost of Clearing

Labor

Cut and pile brush	20 hours
Cut down and remove two large trees	Tree service—get estimate
Relocate part of old stone wall	20 hours

Equipment

Rental of power saw
Rental of truck for removal of brush, etc.
Rental of small bulldozer for ½ day

Overhead

Who carries insurance (liability, workmen's compensation, etc.)?

The labor can be performed by the owner or the owner can hire day labor. Estimates will have to be obtained from the tree service and the equipment rental service, both of which will want to drive their own heavy equipment.

6.4   HOW TO LAY OUT THE BUILDING LOCATION

When the land has been cleared, at least to the extent of giving a clear view of the building site, the corners of the building must be established. This can be done by the use of a surveying transit or by taping. In either case the reference points must be the corner markers of the property, which are either pipe or stakes that have been set in place by a land surveyor. The deed to the land should locate these points by direction and footage.

It is not advisable to use the taping method on larger pieces of land because its accuracy is limited. In such cases the use of the transit is advised not only for location but also for levels. Figure 6.2 shows how taping can be used on a smaller plot. Figure 6.3 shows how batter boards are erected after the corners of the house have been located. The batter boards are used first to locate the approximate line of the excavation and then to establish the exact location of the footings and the foundation wall. Figures 6.4 and 6.5 show how this is accomplished. If a transit is not used, leveling can be done by the use of a water hose with a glass tube at each end. As water seeks its own level, the level of the water in one tube need only be set at the desired height; all other heights will be set by the water level in the other tube (Figures 6.6 and 6.7). It is at this time that the builder should obtain a framing square. This square is simply a large-scale version of a carpenter's square and is made from two pieces of straight wood (1 x 4 pine) that are set at right angles and braced. As shown in Figure 6.8, the dimension can be 3, 4, or 5 ft and 9, 12, or 15 ft. These dimensions will assure the builder of a perfect square.

This section on laying out the exact building location cannot be complete without again mentioning the cautions about the location as mentioned in Section 6.3. If the house is on a slope, it is advisable to raise the foundation by using the highest point of the land as the control point, as shown

Figure 6.2. How to locate the corners of a house by taping (for small plots only).

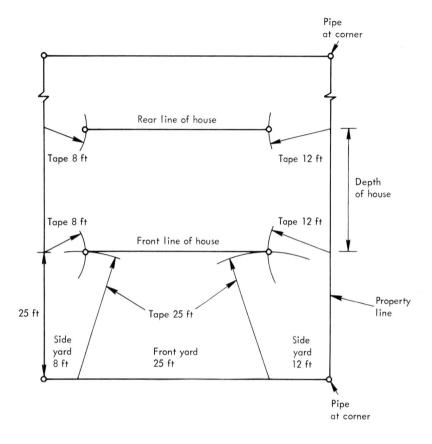

Pipe at corner

Rear line of house

Tape 8 ft

Tape 12 ft

Depth of house

Tape 8 ft

Tape 12 ft

Front line of house

Property line

25 ft

Tape 25 ft

Side yard 8 ft

Front yard 25 ft

Side yard 12 ft

Pipe at corner

Figure 6.3. Establishing of batter boards.

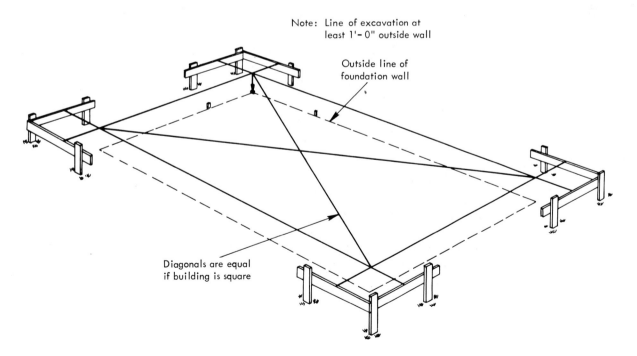

Note: Line of excavation at least 1'-0" outside wall

Outside line of foundation wall

Diagonals are equal if building is square

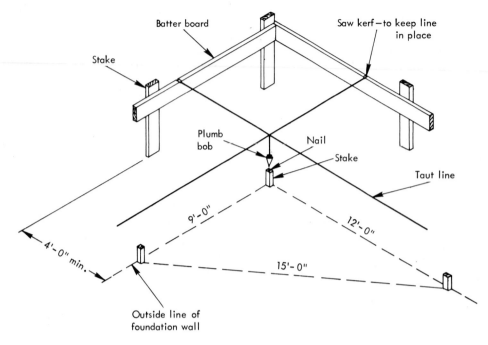

Batter board

Saw kerf—to keep line in place

Stake

Plumb bob

Nail

Stake

Taut line

9'-0"

12'-0"

15'-0"

4'-0" min.

Outside line of foundation wall

Figure 6.4. Detail of use of batter boards.

Figure 6.5. How to use batter boards to locate footings.

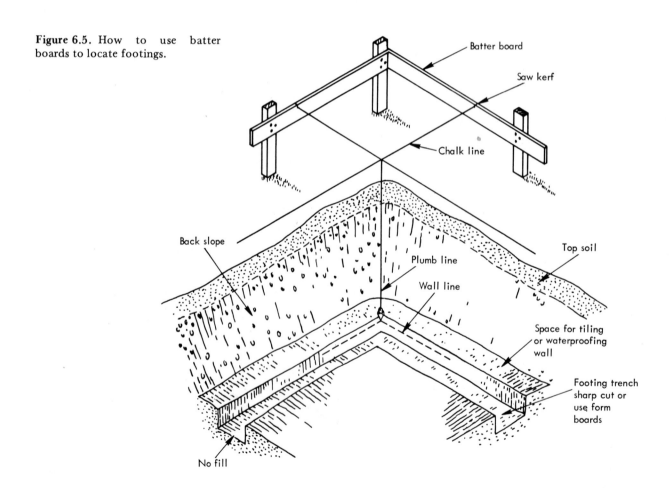

Batter board

Saw kerf

Chalk line

Back slope

Plumb line

Wall line

Top soil

Space for tiling or waterproofing wall

Footing trench sharp cut or use form boards

No fill

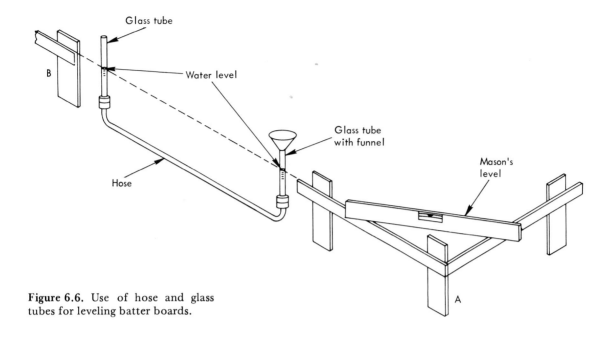

Figure 6.6. Use of hose and glass tubes for leveling batter boards.

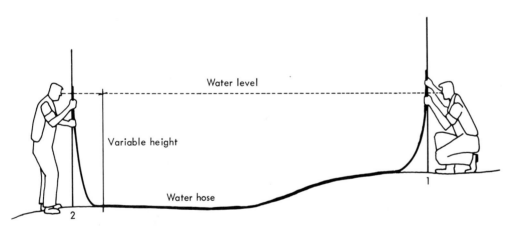

Figure 6.7. Use of hose and glass tubes for any leveling within length of hose.

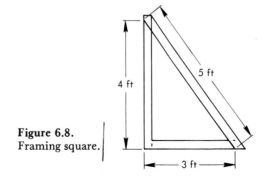

Figure 6.8. Framing square.

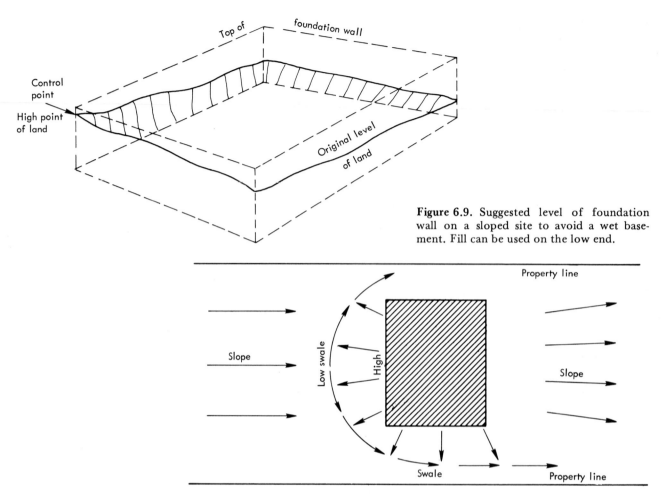

Figure 6.9. Suggested level of foundation wall on a sloped site to avoid a wet basement. Fill can be used on the low end.

Figure 6.10. Locating a house on a slope. Build up soil around the house to shed water down to a lower swale.

Figure 6.11. How to drain low spots on a building site. The foundation should be raised as high as possible.

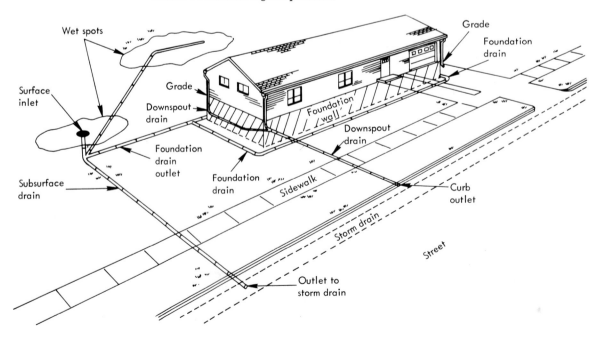

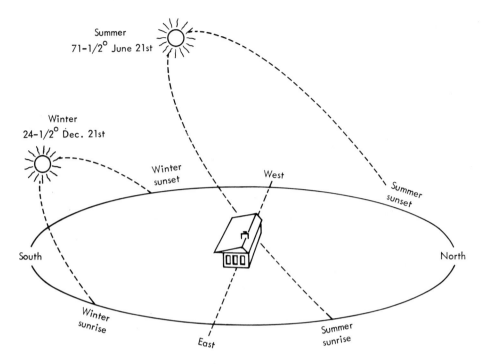

Figure 6.12. Sun direction at 40° North latitude (approximate center of the United States). House or room locations can be oriented by this.

in Figure 6.9, and then to make low points (swales) around the house, as shown in Figure 6.10. If the land is low and there are wet spots during prolonged periods of wet weather, these places can be drained to a lower spot or to a storm sewer, as shown in Figure 6.11.

Figure 6.12 shows sun direction at a point about midway in the United States. If the owner wants morning or afternoon sun in the winter or summer, the house can be oriented so that it gets sun in the living room, bedroom, kitchen, or elsewhere. It can be noted that the winter sun in these latitudes is at a much lower angle than the summer sun. A house can be built with overhangs that shade the summer sun but allows the rays of the winter sun to shine directly into the house.

6.4.1   Who Does the Layout Work and What Will It Cost?

When all decisions have been made as to the general location and orientation of the house and the proper height of the foundation walls with reference to slope and groundwater has been determined, the exact location of the house becomes the next step. The professional builder can often use a transit or certainly has the ability and knowledge to tape the location. The builder can also use a level or the hose technique to determine the proper level for the height of the foundation walls. This can, of course, also be done by the owner-builder, but it must be done very carefully. An error of a single foot may cause a violation of the zoning code and lead to the endless trouble of having to obtain a zoning variance. The erection of the batter boards, which determine the extent of the excavation and the location of the footings, is really a professional job. If there is not a general contractor, the owner is advised to include this location work in the contract of the excavator-foundation builder or the framing carpentry contractor. The taping or location by transit and erection of batter boards and lines should not take more time than that of two skilled laborers for a 1 day. The owner must make sure that the property lines are clearly marked before this work is started. Rates vary, but this should not cost very much and is money well spent. Perhaps the owner, before awarding the excavation-foundation or frame carpentry contract, can ask one of the subcontractors to do the laying out as part of that subcontractor's work.

6.5   PLANNING THE SITE FOR MAXIMUM EFFICIENCY

The prior planning of the construction site (Figure 6.13) should include the following:

For the Excavation

Access to the site by power shovel, ditch diggers, and trucks carrying excavated material out or backfill in.

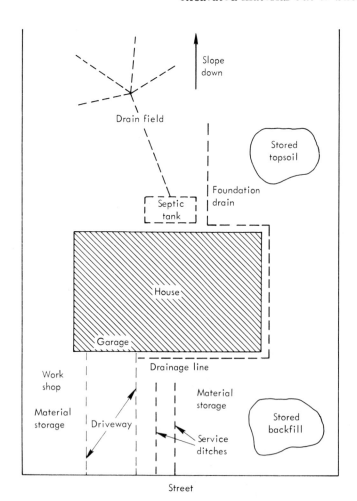

Figure 6.13. Suggested site plan for a small building lot. Septic tank and drain field shown if no sewer is available.

For the Foundations

Access for trucks carrying form lumber, concrete or masonry blocks, gravel, drain tile, cement, and sand.

For Earth Storage

The topsoil, which must be scraped off the site and stored, should first be placed where it does not interfere with the storage of materials or any construction work but where it can be pushed back into place with a bulldozer. The need for backfill should be ascertained before excavation is started. If no fill is needed, the excavated material should be trucked away immediately. Stored earth for backfill, unless there is only a small quantity, is a nuisance on a small site and should be stored as far out of the way as possible. Backfill around the foundation should be put back in place as soon as possible.

For Material Storage

Framing lumber, sheathing, and roofing material should be stored as close to the house as possible and in the order of its use. Some provision must be made to shelter the sheathing and roofing from the weather.

For a Workshop

This usually consists of a power saw and either saw horses or a table. Room should be given so that large sticks of lumber or large panels can be swung around. Location should be between material storage and the house.

For Sanitary Facilities

A portable sanitary facility should be close to the house but in back if possible.

For Septic Tank

If no street sewage line is available, provisions must be made for a septic tank and drain field. This area should be kept clear. A certificate of occupancy will not be granted until this system is connected.

The site plan shown in Figure 6.13 is for a small site. There is, of course, much more freedom of choice on a larger site, but the basic principles of convenience and ease of access remain the same. The builder must never forget that the most expensive portion of any construction job is the labor cost. Any efforts to save steps or to promote accessibility of materials is well worthwhile.

CHAPTER SEVEN

job progress, cost control, and job safety

7.1 THE IMPORTANCE OF A PROGRESS SCHEDULE

One of the most important considerations in the construction process is the preparation of a progress schedule. By preparing such a schedule and staying within its limits, the builder can arrange for the timely deliveries of material, give sufficient lead time to any subcontractors, and arrange for the right workers to be on the job when they are needed. It does not matter how large or small the structure is—a well-thought-out schedule saves time and money. To mention one instance—the interest on a construction loan is much higher than on a permanent mortgage. Why not save this extra interest by completing the work as soon as possible?

The scheduling of construction of a small building can be a comparatively simple job if the builder or owner takes the trouble to think it through and does some investigating. If he or she has had test borings made or test pits dug, it will be known what the excavated material will consist of, and the builder or owner need only seek out the available equipment to remove it.

For the foundation walls the builder should make sure that concrete blocks or concrete can be obtained and, if needed, masons. For framing the structure the builder should make sure that the lumber dealer has all the required sizes of framing lumber and that light steel columns and beams can be obtained when they are needed.

The plumbing and heating and electrical components require that the builder find out when the required kinds and sizes of equipment will be available. The builder must also make sure that the tile, finished flooring, or glass that is called for can be obtained. One job was delayed for 8 weeks waiting for certain bathroom tile; the contractor had never taken the trouble to investigate its availability. This held up final payment and created a great deal of ill-will.

When the progress schedule has been made up, the builder or owner must make every attempt to adhere to it. Constant expediting of subcontractors is a nuisance, but it is important and must be done.

7.1.1 The Bar Chart

The simplest and by far the most widely used progress schedule is the bar chart. Figure 7.1 shows a chart for a residential building. To prepare such a chart the builder must be familiar with the sequence of construction. Following is an explanation of how the chart was prepared.

It should be noted that each trade is given more time than is probably necessary and that the trade lines overlap. This allows for float time and delays caused by weather, delays in delivery, or delay in getting the proper workers on the job.

The excavation starts on January 1. It should be evident almost immediately what type of soil is present and how long the general excavation and footing excavation will take.

The concrete and reinforcing rods (if used) can then be ordered for delivery 2 weeks after the excavation has started.

Foundation wall material is ordered for delivery a few days after the footings are started to allow the concrete to attain its initial set (48 to 72 hours, depending on the temperature). At this time concrete wall forms can be started or concrete blocks and mortar should be on the job ready for the masons.

Figure 7.1. Bar chart for a small house. This progress schedule has enough "float time," or "leeway," so that it can be completed in 4 months, if necessary, by pushing all finish trades into the fourth month.

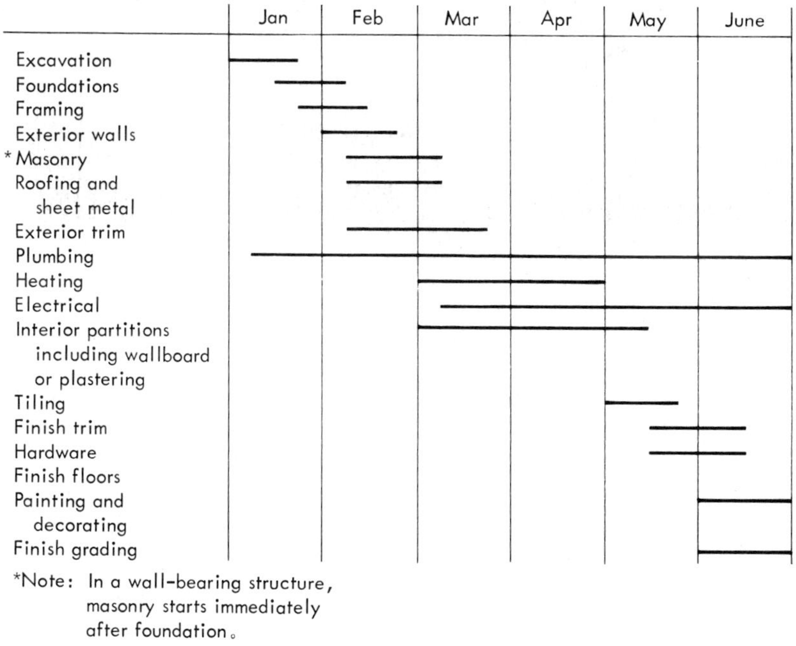

*Note: In a wall-bearing structure, masonry starts immediately after foundation.

Footing drains, if required (in this case they are part of the excavation), can begin to go in as the foundation wall progresses.

Framing lumber is ordered for delivery 1 week after the foundations are started. The lumber can be sorted and the sills can be laid as soon as the concrete or mortar has attained its initial set.

Backfill against the completed foundation wall will be helpful in speeding up the framing. The builder is warned not to pile earth against green concrete or freshly laid blocks. He should wait for at least a week.

Exterior wall material and roofing underlying material should be delivered not more than 1 week or 10 days after the framing is started. The builder should always keep in mind that no matter what the season of the year, the sooner the house is put under cover, the more quickly the job will go.

Masonry for chimneys should be started almost immediately after the exterior walls are started because neither the walls nor the roof can be completed and made watertight until this work is completed.

Exterior windows and frames or other openings should also be ordered at this time. It is best to allow some time for backpainting this material before it is installed.

Plumbing was started almost immediately after the excavation was started, and sewer and water lines have been brought into the basement or crawl-way or into the floor slab forms before the slab is poured. Plumbing continues on and off during the entire construction. Bathtubs and other cumbersome equipment should be brought in and put in place before interior studs are set.

Heating does not need to start until the house is weathertight, but the furnance or other cumbersome equipment should be brought in *and protected* before floor framing is completed.

Electrical work should not be started until the house is watertight and weathertight. Any moisture that can gain access to electrical wiring or equipment can be lethal. No electrical equipment or wiring should be delivered until the house is tight.

All other work follows in sequence as shown. In preparing and using the chart, the builder should be sure to allow lead time for all operations.

7.1.2   The Critical Path Schedule

To prepare and correctly use a critical path schedule, the builder must break down all construction into its component parts, just as would have been done for a bar chart. In the critical path each part of the construction is identified by date. Figure 7.2 shows the start of a critical path diagram. The critical path is rarely, if ever, used in residential construction. It is not even used in large uncomplicated construction such as simple office buildings. The high degree of construction and technical sophistication required for a successful critical path schedule is useful in the construction of complicated processing plants (power plants, refineries, chemical plants, bridges, etc.). Figure 7.3 shows an actual bar chart used in the construction of a 34-story office building which was *completed on time.*

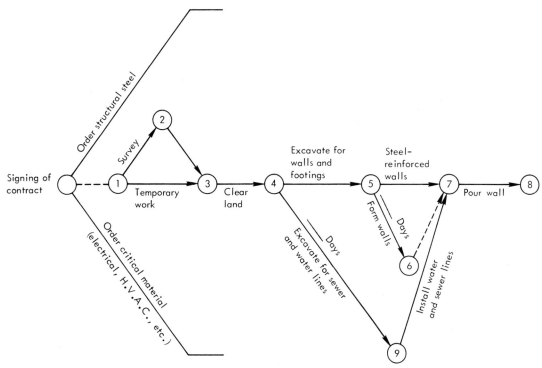

Figure 7.2. Simplified example of a critical path network diagram. The time allowed for each activity would be placed on the diagram at each arrow.

Figure 7.3. Actual bar chart progress schedule for a 34-story office building with three basement levels. Tenants started to move in 2 months before final completion (which was on time).

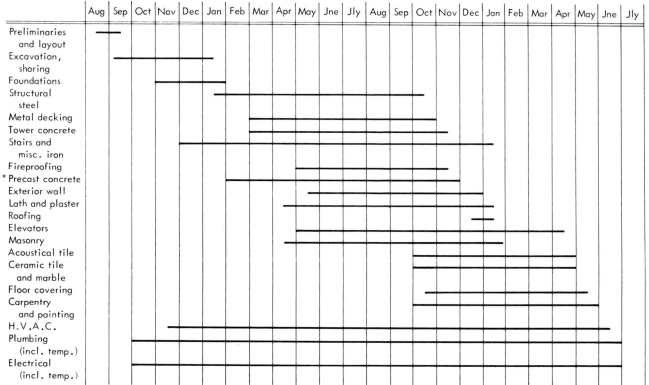

* Precast concrete was used on the exterior walls.

7.2   COST CONTROL

The builder or owner-builder who is performing a job for which a fixed sum has been allotted, either by bid or by the use of fixed amounts of construction payouts, is obviously under extreme pressure to keep costs within these fixed limits. There are three principal items of cost in any construction job. First is overhead, which can be divided into general overhead, which in a professional builder's case may be called the cost of doing business, and overhead directly attributable to the specific project. Second is the cost of material, in which can be included equipment rental and other equipment costs. Third is the cost of on-site labor, which should include all payments by the employer of Social Security taxes, any union benefit taxes, unemployment insurance taxes, and any other costs directly attributable to payroll.

7.2.1   Overhead Control

General Overhead

Chapter 4 included a comprehensive list of a builder's overhead expense. In estimating any job, the builder must allot a certain portion of this overhead to the particular job. One way of doing this is to estimate the length of the job and then apportion the overhead accordingly. For instance, if there is an overhead of $3600 per year (not including any personal expenses or pay for the builder) and the builder constructs six houses a year, each house (if each is approximately the same size), can be charged $600 for general overhead. Or the builder can weight the overhead charge by the total annual gross business income. For instance, for $300,000 per year of business, each $1000 would carry $12 of general overhead.

To control this, the builder should keep a monthly total of actual expense on an accrued basis and compare it to estimated expenses for the year. The builder can then see immediately what is happening and why, and has time to correct it. It would not seem very important for a small builder with no office payroll and low overhead to go through this exercise, but it does not take much time and is a very good habit to get into. It is the first step in cost control.

Job Overhead

As the term implies, job overhead consists of the direct expense sustained by the specific construction job. This expense is sustained by anyone who builds. The items are:

Insurance—workmen's compensation, public and automobile liability, and any other insurance (including fire and extended coverage).
Permits and licenses (e.g., street paving, sewer, and water).
Cost of outside supervision.
Interest on advances during construction—owner pays.
Taxes during construction—owner pays.
Job telephone, sanitary facilities, and temporary water and power.

Many of these costs are fixed by public authorities or by utilities, but the builder can control them by shortening the job—and certainly can control the costly item of insurance by job safety.

7.2.2 Material Control

There are four basic principles that must be observed for proper material control: the builder must make sure (1) that the quantity of material delivered checks with the delivery ticket, (2) that the material is of the quality ordered, (3) that the quantity of material delivered checks with the quantity that is installed (except for visible waste), and (4) that the material is guarded against outside pilferage or malicious mischief.

Quantity and Quality Checks

Every delivery of material should be accompanied by a delivery ticket, which should spell out the quantity and quality of the material (Figure 7.4). The builder or a trusted worker should make it a point to be present when large quantities of material are delivered to make sure that the material is placed in the proper location and to check quantity and quality as soon as possible after delivery. Any claims for irregularities should be made to the material dealer as soon as possible.

Material Used

The builder can keep control not only of the material but also of the labor cost by a simple check of the material delivered, the material that has been erected, and the material that is left. Figure 7.5 shows a simple form that can be used for this purpose. The quantity of material used tells

Figure 7.4. Sample of a delivery slip.

JONES LUMBER COMPANY						
Sold To _____			Date Sold _____			
_____			Date Delivered _____			
Job or Order Number _____			Sold by _____		Driver	
Code	Qty	Description	Units	Price	Amount	
Delivered prices based on Tailboard Delivery. Please examine goods before accepting. Our responsibility ends after acceptance.			Subtotal			
			Tax			
			Total			

MATERIAL DELIVERED AND USED

BUILDING *J. R. Smith House – Allwood Road*

DATE RECD.	DEALER	MATERIAL BY DEL. TICKET	DATE COUNTED	MATERIAL ERECTED	MATERIAL ON SITE	REMARKS
1/1/7-	Tuttle Lumber	100 2x4 – 8' 33 - 2x4 – 12'	1/2/7-	85 - 8' 27 - 12'	10 - 8' 6 - 12'	Only waste on 8' is to cut down to 7'-6". Where are 5-8' sticks?

Figure 7.5. Form for checking material delivered, used, and on site.

MATERIAL TRANSFER

BUILDING *J. R. Smith House*

DATE	MATERIAL	QUANTITY	PRICE	TRANSFERRED TO
1/7/7-	2x4 – 8'	10	See Unit	Perkins House

Figure 7.6.
Form for recording transfer of material from one job to another.

the builder whether he or she is getting a full day's work from all personnel, and the amount left over indicates whether there has been excessive waste or pilferage.

Sometimes a builder will transfer material from one job to another. It is wise to keep an account of this. Figure 7.6 shows a simple form that can be filled out in a few minutes. This enables the builder to charge the material to the job in which it was used.

Pilferage and Malicious Mischief

When lumber and other usable materials are stored in the open, as they must be during part of the construction, the problem of pilferage can become a serious problem in some localities. Easily portable material can be locked in a sturdy shanty, and bulky lumber, brick and so on, can be covered by tarpaulins to make it less visible. Local police can be asked to have a patrol car check at intervals. It is possible to cover pilferage and malicious mischief by a rider on the builder's liability insurance. A daily check can alert the builder to any loss.

7.2.3 Labor Cost Control

In estimating the cost of construction, the carpentry contractor, the masonry contractor, the roofer, or any other subcontractor should know within a close approximation how many feet of stud wall a crew of two can erect, how many concrete blocks a mason and tender can lay, or how many square feet of roof a roofer and tender can lay. This is especially true if the subcontractor knows the workers. Unfortunately, however, this is not always true, and a strange crew may not be able to (or want to) match its production to the contractor's estimate. There are several things a contractor can do to maintain production. First, the contractor should make sure that all necessary material is on hand when it is required. For instance, the arrival of sheathing before the framing is completed, or the concrete truck arriving just as the one on the job is unloading, tends to keep craftsmen busy. Second, the contractor can use one of the steady workers as a "pusher." This worker can be notching girders as the framing is being completed or see to it that the mortar mixer produces a fresh batch before the old batch is completely used. Third, the contractor should keep some kind of a daily progress report in some detail (Figure 7.7). With such a report he can match actual progress to the estimate. Such a report would supplement the information in the chart shown in Figure 7.5.

DAILY PROGRESS REPORT

Figure 7.7.
Daily progress report.

Project		Job #	Date
Weather		Temperature	
Supt. or Foreman			

Journeymen on Job	No.	Total Hours	Work Completed
Mason	1	7	Chimney foundation
Laborer	1	7	"
Carpenters	2	14	Completed all rafters

Subcontractors on Job

Plumber	1	7	Plumbing roughing 1st fl.
Helper	1	7	"

Equipment on Job

Bulldozer filling around foundation walls. 50%

Change Orders, Special Instructions, Accidents, Remarks

7.2.4 Change Orders

One of the most financially troublesome items that a builder encounters during construction is a change in plans or specifications. Such changes can come after the work as called for by the original plans and specifications has been partially or even fully completed. The tearing down and rebuilding is time-consuming; it can be very costly; and with few exceptions it can cause serious financial disagreement among the builder, the owner, and the architect. Even when the builder is the owner, there are always subcontractors to be considered.

The only certain way for the builder to protect against this is to reduce the entire operation to writing so that the owner knows what it will cost before the work is started. The procedure to be followed is as follows:

As soon as the owner, architect, or authorized agent orders a change, the builder should stop any work that will add to the work that is to be changed.

The builder must ascertain *exactly* what the change is to consist of by getting a sketch and a description, which *should be initialed.*

The builder must then prepare an estimate of the cost of the change (not forgetting to charge for the delay caused by the change) and must set a *time limit* on when the estimate must be accepted or rejected (in writing) and the time to be allowed for the change to be made.

CHANGE ORDER NO. _1_   DATE _3/5/7-_

PROJECT _J. R. Smith House_   CONTRACTOR _ABC Construction_

Figure 7.8.
Change order form.

You are hereby authorized to proceed with the work as described below.

Drawings _Drawing showing change in location doorway L.R. to D.R. - dated 3/2/7-_

Correspondence _None - Verbal discussion_

Work to be Completed by _____

DESCRIPTION OF CHANGE	ADD	DEDUCT	COST
Eliminate existing opening. Frame new one. New wallboard, new trim. Reuse door. Labor $150. Material $50.	200		
TOTALS	200	—	$200
ADD OR DEDUCT FROM CONTRACT	$200		$200

SIGNED _John R. Smith_
Owner or Authorized Representative

Change orders are always trouble. Figure 7.8 shows a simple form that can be prepared by the builder and signed by the owner or an authorized representative.

7.3 JOB SAFETY

Construction is a hazardous business at best and as such the cost of insurance against these hazards makes up a considerable portion of the builder's cost. Actual experience has shown that any serious accident to a worker on a construction project causes a slowdown in the work. The practice of safety on a job therefore not only prevents injury to workers but makes good economic sense to the builder. In this connection the builder is again warned that injury to workers must be reported immediately on forms prescribed by the workmen's compensation authority.

7.3.1 Protection of the Worker

Workmen's compensation insurance premiums are based on an experience rating which can increase or decrease an employer's payments based on the past accident record. Some simple precautions against job accidents are as follows:

Avoid having personnel work in deep and narrow trenches unless the sides are shored.

Keep areas cleared of boards with protruding nails. Ask workers to wear heavy-soled shoes.

Protect holes in the floors.

Avoid unsafe ladders with cracked rungs or side rails. Check scaffolding for weak planks or supports.

Check ropes or wire supports for chafing.

Do not allow top-heavy piles of material.

Be sure that concrete or mortared load-bearing walls have attained their proper set before any load is placed on them.

Very important—keep the work area clear of accumulated rubbish. This is one of the basic rules for speeding the work and avoiding accidents.

7.3.2 Protection of the Public

Some simple precautions can save the builder law suits and premium dollars.

Don't pile material too close to a public walk.

Don't leave holes or rubbish on a public walk.

When a hose or cable crosses a public way, it should be bridged over or otherwise protected.

Any large pile of topsoil, fill, sand, and so on, should be left at a reasonable angle so that there is no danger of cave-ins.

Advise all workers to watch carefully before crossing a public walk in a motor vehicle or with any equipment.

Carefully lock all power equipment.

7.3.3   Fire Protection

The first and most important precaution against fire is to keep the work site clear of accumulated rubbish. If the area is kept clean, a carelessly discarded cigarette or a spark from a torch will have nothing with which to start a fire.

Space heaters are very dangerous; if they must be used overnight or when no one is on the job, they should be adequately protected and kept as far away from flammable material as possible.

Frayed or otherwise insecure electric wiring should not be allowed on a job. In any case, the main switch should be pulled and the switch box locked before the job is left.

The builder or a trusted employee should go through the building every day after work is completed to look for any signs of smoke or any other indication of an incipient fire.

7.3.4   Some OSHA Requirements

The Occupational Safety and Health Administration was created by the Congress several years ago. Its duty is to protect workers against hazardous working conditions. OSHA inspectors have had the power to stop a construction job if the builder was in serious violation of safety standards. Since the OSHA regulations were established, many of the petty details and costly requirements have been eliminated. The regulations for construction safety confine themselves to the protection of stairwells and holes in floors and walls, the wearing of hard hats, the safety of ladders and scaffolds, job sanitation, and several other safety measures that any careful builder would take in any case.

CHAPTER EIGHT

excavation and foundations

8.1 TOPOGRAPHY AND SOILS

Excavation for the footings, basement, and foundations for a residence can consist of bringing in a power shovel and trucking away the excavated material. Or it can consist of blasting ledge rock or digging on a slope where fill will be required or excavating to a level below the water table. It is strongly recommended that before any digging is started, one or two test pits be dug to ascertain the kind of soil that is below the surface, to find if there is ledge rock, and to see how deep the pit can be before water seepage occurs. The builder may very well decide to either eliminate or greatly restrict a basement if there are strong signs of high groundwater, an underground stream, or ledge rock. Waterproofing a basement against a water table of more than 2 or 3 ft above the basement floor *in the wet season* can be very expensive. If the water table is this high in the dry season, then serious thought should be given to either a crawlway only or a flat slab.

The character of the soil is also important. Sandy soil will be good for water drainage but may require larger footings and will not ordinarily sustain steep slopes, so that shoring may be necessary for deep holes. Clay soil can be cut sharply but is poor for drainage. The test pits will show what may be expected. The builder should watch for land fill. Large areas of soft leafmold or signs of rotted wood or other debris indicate that the footings should be taken down to solid undisturbed soil.

8.2 EXCAVATION

Before the builder calculates the amount of material to be excavated, several factors must be taken into account. The first is how much clear head room

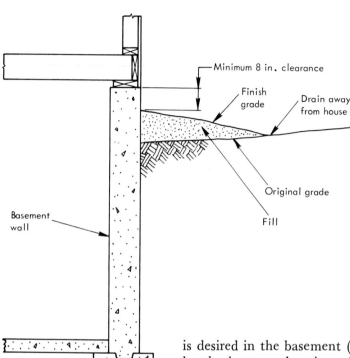

Minimum 8 in. clearance

Finish grade

Drain away from house

Original grade

Fill

Basement wall

Figure 8.1. Use of fill to drain ground water to a low point.

is desired in the basement (or how deep the crawlway is to be or what will be the bottom elevation of the floor slab). For the purpose of this book, the calculations will be shown for a full basement. The second factor is how high above the ground the foundation wall is to be. Foundation walls can have a full height of anywhere from 7 ft 4 in. to 8 ft. They should extend above finished grade around the house by enough height so that wood members can be protected from termites or carpenter ants by means of periodic inspections of the bare foundation wall. In the case of a crawlway, a clear height of at least 3 ft 6 in. is recommended so that inspections can be made for termites or repairs can be made to heating or plumbing or electric work. The wood members of a house built on a slab must also be protected and the builder is advised to raise the slab so as to show at least 1 ft of clear foundation wall. The third factor, which really has to do with the height of the walls, is the matter of drainage. The foundation wall should be high enough so that the finished grade can slope down from it to shed water away from the house (Figure 8.1).

8.2.1   Calculation of Quantities

Excavated material is measured by the cubic yard, which is 27 ft³ (3 ft x 3 ft x 3 ft). On a flat site or one with a slight slope (not more than 1 ft in 20 in any direction), the simplest way is to multiply the width by the length by the depth of the excavation, as shown in Figure 8.2. In this instance the outside dimension of the foundation walls is 25 ft by 40 ft and the excavation is 7 ft 6 in. deep. The amount of earth to be moved depends on the nature of the soil. If the soil is reasonably firm and can hold a bank at a sharp angle, the excavation can be 2 ft wider than the outside of the walls at the bottom (to allow for parging, footing drains, etc.) and slope sharply upward to 3 ft 6 in. wider at the top. The width of the excavation would then be 25 ft plus the average between 2 ft at bottom and 3 ft 6 in. at the

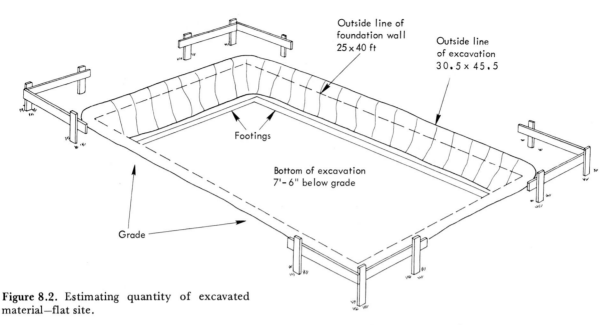

Figure 8.2. Estimating quantity of excavated material—flat site.

top or 2 ft 9 in. This makes a total of 25 ft plus twice 2 ft 9 in. or 30 ft 6 in. The length would be 40 ft plus 5 ft 6 in. or 45 ft 6 in.:

$$30.5 \text{ ft by } 45.5 \text{ ft by } 7.5 \text{ ft} = 10,408 \text{ ft}^3 \text{ or } 385 \text{ yd}^3$$

The calculation for a sloped site can be made as shown in Figure 8.3. The highest point of the site should be taken as the control point. In this case it can be designated as $0'$-$0''$ elevation. The ground slopes to $-3'$-$0''$, $-4'$-$0''$, and $-5'$-$0''$ at the other three corners. The average of the 4 corners is $3 + 4 + 5 + 0 = 12 \div 4 = -3'$-$0''$, which is the average height of grade. The foundation wall is $7'$-$6''$ high and should project $1'$-$0''$ above grade at the highest point or an elevation of $+1'$-$0''$. The bottom of the excavation is $7'$-$6''$ below this, or $-6'$-$6''$. As the average height of grade is $-3'$-$0''$, there will only be $3'$-$6''$ of earth to remove. The total amount then is

$$30.5 \text{ ft by } 45.5 \text{ ft by } 3.5 \text{ ft} = 4857 \text{ ft}^3 \text{ or } 180 \text{ yd}^3$$

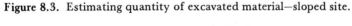

Figure 8.3. Estimating quantity of excavated material—sloped site.

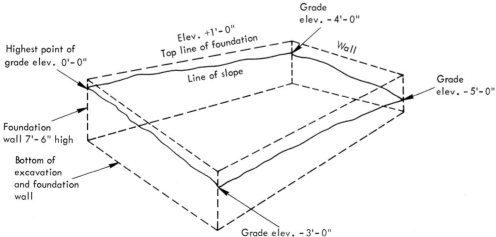

Such a sloped site lends itself to a split-level house. With the basement floor at the same level, the house can be built to provide above-grade living space with a basement under part of the house and a crawlway under the lower portion. The exterior grade level will then have to be cut down at this end so that the lower portion of the house will be above grade. If the builder does not want a split-level house, the space can be used for a basement room or a garage.

8.2.2 Footings

Footing trenches can be dug by hand or by machine and can very often be made just wide enough and deep enough so that concrete can be poured directly into the trench. If the soil is not firm enough to allow this, the trench must be made with sloping sides and forms must be built to confine the concrete to the exact width and depth as shown on the plans. Figures 8.4 and 8.5 show methods of building and bracing these forms. In many cases, where the soil corresponds to No. 6, 7, or 8 in Table 2.2, page 25, it could cost less for the builder to excavate to footing bottom, form the footings as shown in Figure 8.4, and then fill back up to basement level. All of this can be a machine operation.

In cases where the soil is unstable and does not meet the requirement for its bearing value (Table 2.2), the builder may have to drive some piles to reach soil of the proper bearing value. Pile driving is very expensive, however, and if only a few feet is involved in getting down to good bearing soil, the builder may prefer to dig down and build up to basement grade with masonry blocks on a concrete footing or he may use poured concrete columns (which is an expensive type of construction for a private residence). In such a case, the weight of the foundation wall and the structure is carried by a grade beam, which spans between the columns at footing level and acts as the footing (Figure 8.6). The size and spacing of the columns and the size of the grade beam can be determined by a simple engineering calculation. The builder is advised to employ the services of a structural engineer, who will probably have to initial a drawing to satisfy building authorities.

The size of a footing for the most frequently encountered bearing soil (Nos. 4, 5, 6, 7, and 8) is shown in Figure 8.7. An 8-in. block wall will take a footing that is 8 + 4 + 4 = 16 in. wide by 8 in. deep. A 10-in. concrete wall will take a footing 20 in. wide by 10 in. deep. In the case of a concrete wall, the footing should be keyed as shown by the dotted lines. Such

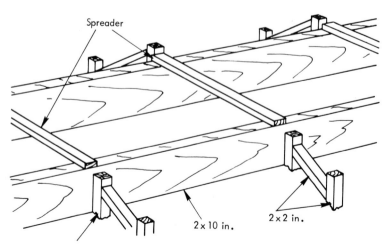

Figure 8.4. Typical wall footing form, no footing trench.

Spreader

2 x 10 in.

2 x 2 in.

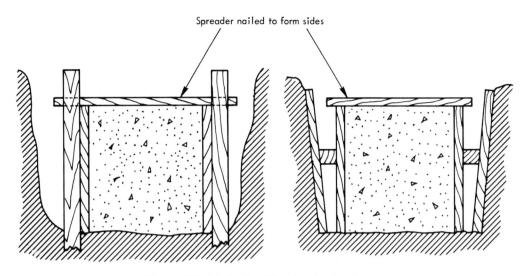

Figure 8.5. Methods of bracing footing forms.

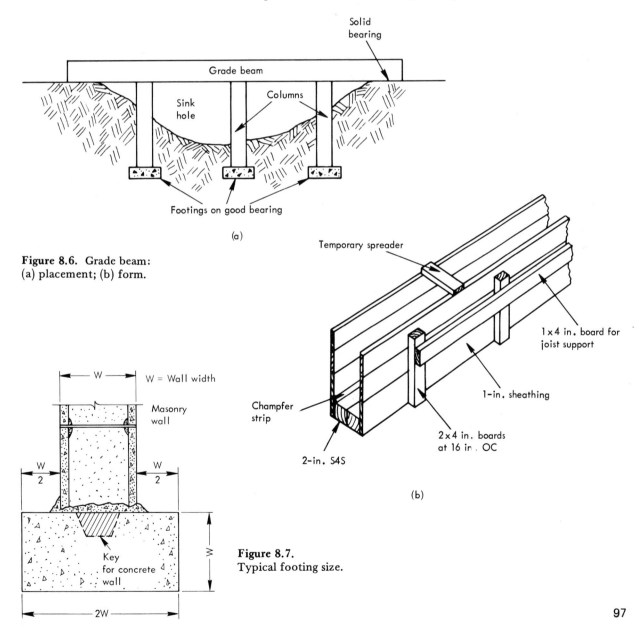

Figure 8.6. Grade beam:
(a) placement; (b) form.

Figure 8.7.
Typical footing size.

97

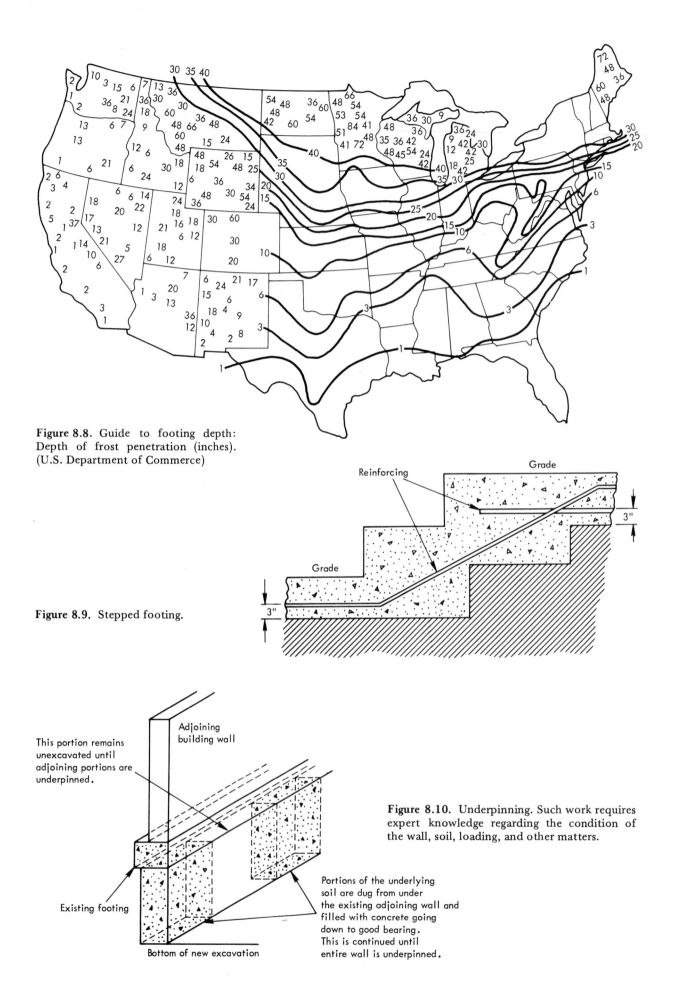

Figure 8.8. Guide to footing depth: Depth of frost penetration (inches). (U.S. Department of Commerce)

Figure 8.9. Stepped footing.

Reinforcing

Grade

Grade

3"

3"

This portion remains unexcavated until adjoining portions are underpinned.

Adjoining building wall

Existing footing

Bottom of new excavation

Portions of the underlying soil are dug from under the existing adjoining wall and filled with concrete going down to good bearing. This is continued until entire wall is underpinned.

Figure 8.10. Underpinning. Such work requires expert knowledge regarding the condition of the wall, soil, loading, and other matters.

a footing on soil as mentioned above should be adequate for the normal 2- to 2½-story frame structure.

Figure 8.8 shows the frost-penetration depths for the continental United States.

Stepped Footings, Underpinning, Dewatering, and Footing Drains

There are two ways of constructing footings on sloped ground. The obvious way, if a basement is planned and if the slope is not too great, is to excavate so that all footings are on the same level below grade, as shown in Figure 8.3. If a full basement is not desired, the footings can follow the slope of the hill. Such a footing is known as a stepped footing. Figure 8.9 shows such a footing. The depth and width of the steps depend on the kind of soil and the loading on the footings. The bottom of the footing must be on solid undisturbed ground. The vertical step should not exceed 2 ft. The length of the horizontal step should be at least 1.75 times the vertical height. The slope should not exceed a ratio of 2:1 and it is advised to place a pair of No. 4 (½-in.) reinforcing rods as shown in Figure 8.9.

The builder of the ordinary private residence does not usually encounter a problem with neighboring foundations. However, there may be instances in densely populated areas where a new foundation may come so close to an existing foundation that there is danger of settlement. If the new foundation goes down only to the same level as the bottom of the existing footings or if it goes deeper and it is far enough from the existing footings for a satisfactory angle of repose (which should be determined by a soil test), no underpinning of the existing foundations is necessary. But if the soil is unstable or for some reason the new foundation must go much deeper than the neighboring existing foundation, underpinning is necessary. Figure 8.10 shows how underpinning is accomplished.

The dewatering of an excavation is a very expensive operation and is not justifiable for any but a very large house, and then only if a basement below groundwater level cannot be avoided. Casual water in an open excavation may be controlled by a pump, but if the water level at the footings is higher than the bottom of the footings, the poured concrete of the footings must be kept dry until the concrete has attained at least its initial set. This can best be done by well points, as shown in Figure 8.11.

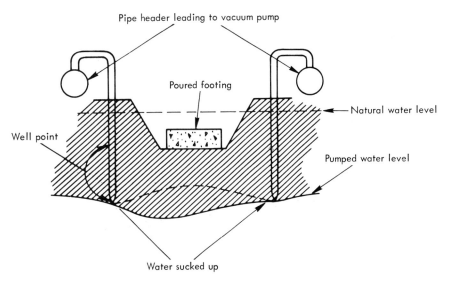

Figure 8.11. Well-point installation. Well points lower the water level until the footings are poured and set.

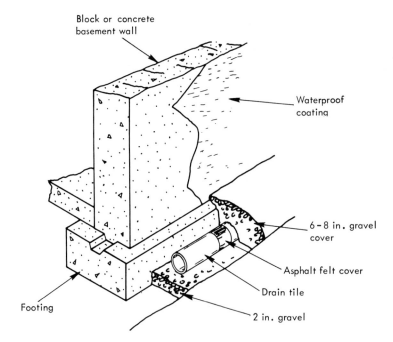

Block or concrete
basement wall

Waterproof
coating

6-8 in. gravel
cover

Asphalt felt cover

Drain tile

Footing

2 in. gravel

Figure 8.12.
Use of drain tile at footings.

Footing drains are always advisable around foundation walls that enclose a habitable basement area. The installation is not expensive and can be done immediately after the foundation walls are started (if they are masonry) or immediately after the forms are stripped (if concrete). Clay drain tile or perforated transite or other noncrushable perforated pipe is laid on a bed of crushed stone below the level of the basement floor and is covered with at least 6 in. of crushed stone, as shown in Figure 8.12. The water must be led to a level lower than the bottom of the foundations or to a dry well or sump.

8.2.3 Cost Estimating for Excavation

The first step in estimating the cost of any kind of construction is to determine the quantities involved and then to estimate the unit cost of the kind of work to. be done. In excavation the builder will encounter the general digging of the basement hole, which will undoubtedly be done by machine, and the digging of holes for pier footings and trenches for foundation wall footings, which is best done by hand labor. The federal government has made time studies of machine and hand labor in excavation and Tables 8.2 and 8.3 show the results of these studies.

Note: The following costs are as of 1979-1980. Construction costs (as all other costs) are on a steep uptrend. These figures are nationwide averages. Local costs may vary considerably.

Table 8.1 shows the average output in cubic yards per hour of various excavating machines. For an example, a backhoe in common earth will excavate 45 yd³ per hour and load it into a truck or into a pile or leave it for a bulldozer to push away. If a backhoe and operator costs $45 per hour, the cost per cubic yard will be $45 ÷ 45 = $1. The cost of hauling (depending on distances) will be at least the same. The total cost per cubic yard

TABLE 8.1. Average Output of Power Equipment in General Excavation

Equipment	Type of Material	Average Output (yd³/hr)
Power shovel (½ yd³ capacity)	Sandy loam	70
	Common earth	60
	Hard clay	45
	Wet clay	25
Short-boom dragline (½ yd³ capacity)	Sandy loam	65
	Common earth	50
	Hard clay	40
	Wet clay	20
Backhoe (⅓ yd³ capacity)	Sandy loam	55
	Common earth	45
	Hard clay	35
	Wet clay	25

will then be $2 for common earth with no large boulders, no ledge rock, and no water.

The cost goes up dramatically when hand labor is involved. For footings and pier holes below basement grade, Table 8.2 shows that general hand excavation in sandy silt-clay (which may be assumed equivalent to common earth) will take 1 labor-hour for 1.2 yd³ of material. Labor now averages about $9 per hour, including benefits. $9. ÷ 1.2 equals $7.50 per yard. Trench excavation is much slower, and for this the builder should allow $20 per yard, then add $4.50 for loading the material on a truck, for a total of $24.50 per cubic yard. If we take the 25- by 40-ft footing shown in Figure 8.2, the quantity involved would be:

Footing 16 in. wide by 8 in. deep by 130 ft long:

$$1.33 \times 0.66 \times 130 = 114 \text{ ft}^3 \div 27 = 4.2 \text{ yd}^3 \times 24.50 = \$102.90$$

TABLE 8.2. Average Output of Hand Labor in General Excavation

Type of Material	Cubic Yards per Worker-Hour					
	Excavation with Pick and Shovel to Depth Indicated				Loosening Earth—Worker with Pick	Loading in Trucks or Wagons—One Worker with Shovel and Loose Soil
	0–3 ft	0–6 ft	0–8 ft	0–10 ft		
Sand	2.0	1.8	1.4	1.3	—	1.8
Silty sand	1.9	1.6	1.3	1.2	6.0	2.4
Gravel, loose	1.5	1.3	1.1	1.0	—	1.7
Sandy silt-clay	1.2	1.2	1.0	0.9	4.0	2.0
Light clay	0.9	0.7	0.6	0.7	1.9	1.7
Dry clay	0.6	0.6	0.5	0.5	1.4	1.7
Wet clay	0.5	0.4	0.4	0.4	1.2	1.2
Hardpan	0.4	0.4	0.4	0.3	1.4	1.7

TABLE 8.3. Summary of Excavation Costs

	Unit Labor	Unit Material	Equipment	Total Unit Cost	Total Units	Total Cost
Survey and batter boards						
General excavation						
Allowance for boulders, ledge rock						
Allowance for dewatering						
Move and pile topsoil						
Move and pile backfill						
Footing excavation						
Trench excavation						
Excavation for miscellaneous footings (chimneys, areaways)						
Deep footings or piles						
Backfilling, rough grade						
Finish grade, topsoil						
Seed, plant shrubs, etc.						

According to these figures, it would cost $103 to dig the footing and load the material on a dump truck. The productivity of labor and the uncertainties of weather and other factors can seriously affect this cost, so that a substantial contingency should be added.

As mentioned above, none of the quoted costs take account of excavation difficulties, ledge rock, the handling of large boulders, poor bearing soil, or soil that requires shoring. Groundwater and excavating on hillsides will add substantially to the cost. There is no way of estimating this cost until the particular site has been carefully inspected, *and the reader must again be cautioned that the prices mentioned must be carefully checked.*

A summary of the various operations that may be involved in excavation is given in Table 8.3. The estimator is reminded of the caution that is necessary in any construction cost estimating, as mentioned in Chapter 4. The prices quoted in this section are examples only. Prices and labor productivity vary widely.

8.3   FOUNDATIONS

The materials most frequently used for foundations are concrete masonry blocks and poured concrete. These materials are approved by all building codes and are used for houses with full or partial basements, with crawlways, and for those which are built on flat slabs. There are other foundation materials, such as treated timber and rubble stone, but in almost all cases the use of these requires special permission from the building authorities.

8.3.1   Concrete Mixtures: Strength and How to Use

Concrete is a mixture of sand, cement, and aggregate (crushed stone or gravel) to which is added enough water to make it plastic. The sand (fine aggregate), portland cement, and the crushed stone or gravel (coarse aggregate) are thoroughly mixed in certain proportions which determine the strength and plasticity, and then water is added in a carefully measured amount. The water/cement ratio (gallons of water per 94-lb bag of cement) is the determining factor in the final strength of the concrete.

The water used for concrete should be free of all alkalies, acids, oil, or organic matter. Water used for concrete should be fit to drink. In some areas the water may contain an excessive amount of sulfate. This may be fit to drink but should not be used for concrete. The aggregates can be moist but not wet.

The cement most often used is portland cement of Type I or III, which for ordinary use need not be "air-entrained." Air-entrained concrete contains minute air bubbles throughout its content. The air bubbles are caused by a chemical which can be mixed with the portland cement. Air-entrained concrete should be used in all concrete exposed to freezing, thawing, and for all paving. It will also prevent chipping and scaling. The air bubbles in the concrete allow freezing water in the concrete to expand.

For the ordinary 2½-story frame house with a wood or brick veneer or even solid masonry exterior, the typical code and specification will call for 2500- to 3000-psi concrete. This means a mixture to produce concrete that will attain a compressive strength of 2500 to 3000 pounds per square inch after 28 days.

Without going into a great deal of design detail, a simple guide for attaining this strength is as follows. For a strength of 2500 psi after 28 days, the mixture is 1 part portland cement, 2 parts sand, and 4 parts stone by weight to be dry-mixed, after which 7½ gal of water is added. If only 5½ gal of water is added per bag of cement, the strength will go up to 3200 psi. If the concrete is to be mixed by the builder, the simplest way to do this is by volume. For instance, for 3000-psi concrete, the mixtures are as follows:

	Cubic Feet	Weight
6.2 bags of cement at 94 lb per sack	2.97	582
34 gal of water at 8.3 lb per gal, divided by 62.4 gal/ft^3	4.54	283
1758 lb of coarse aggregate	10.80	1758
Air is automatically entrained during the mixing. It amounts to	1.62	
1170 lb of clean sand	7.07	1170
	27.00	3793

This mixture will weigh approximately 3793 lb/yd^3 or 140 lb/ft^3.

In most instances, concrete can be obtained from so-called "ready-mix" companies, which deliver the concrete in trucks that use mounted revolving drums to do the mixing on the job. These companies will deliver at a distance of at least 10 miles at a standard price. There are such companies in large towns almost everywhere in the country. If such concrete is not available, it must be mixed by machine or by hand on the job. Hand mixing is only for footings. A motor-driven batch mixer should be used if the foundation wall is to be of concrete. A wall such as shown in Figure 8.3 will take about 25 yd^3 of concrete and the basement floor will take another 20 yd^3. Figure 8.13 shows a batch mixer which is easily transportable and is satisfactory for small jobs. These units come in sizes of from 3½ to 9 ft^3 capacity. If 8-minute loading time and 2-minute mixing time are allowed, this unit will produce at least five batches, or 45 ft³ (1.6 yd³) per hour. In a 7-hour day if the builder uses this machine for a 20-yd (or more) pour, a bulkhead must be built in the forms, as shown in Figure 8.14. Concrete must never be allowed to stand more than 45 minutes before another batch is added, and this must be kept up until the section of wall is completed.

Larger mixers are available, of course, but they are not easily transportable and are too expensive for the average small builder.

When pouring concrete, the builder should be very careful to deposit it in the form in such a way as to prevent the separation of the various ingredients. This will guarantee maximum strength and minimum honeycombing. Some rules are:

The maximum drop should not be more than 5 ft unless a chute is used.

The ends of the wall should be poured first, at least for the first layer. This is to prevent excess moisture from gathering there.

The concrete should be poured in horizontal layers of not more than 20 in. except in reinforced concrete, where the layers should not exceed 10 in.

Concrete must *not* be allowed to attain its initial set until the entire wall is completed. If this cannot be done, it will be necessary to place keyed bulkheads where the pour has stopped (Figure 8.14).

The concrete should be rodded to avoid excessive honeycombing.

Concrete should be left approximately in the place where it is originally deposited. Too much movement tends to separate the ingredients.

As a conclusion, the builder is reminded to position the hold-down bolts before the pour is started (Figure 8.15). Concrete should not be put in place near 40°F if the temperature is falling. The concrete must be protected against freezing by using warm water, warm aggregates, and air-entraining additive, and then covering the work with salt hay or other protective layering material, and by using high-early-strength additives as well. He must protect concrete against quick loss of moisture, especially on large flat surfaces by covering with tarpaulin, wet salt hay, or wet burlap.

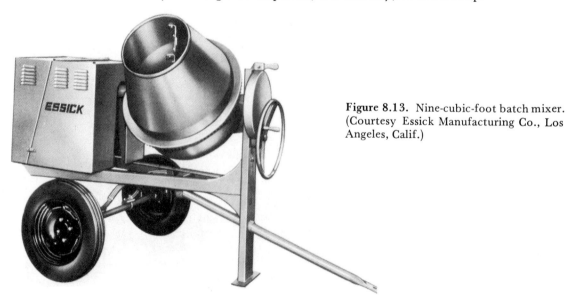

Figure 8.13. Nine-cubic-foot batch mixer. (Courtesy Essick Manufacturing Co., Los Angeles, Calif.)

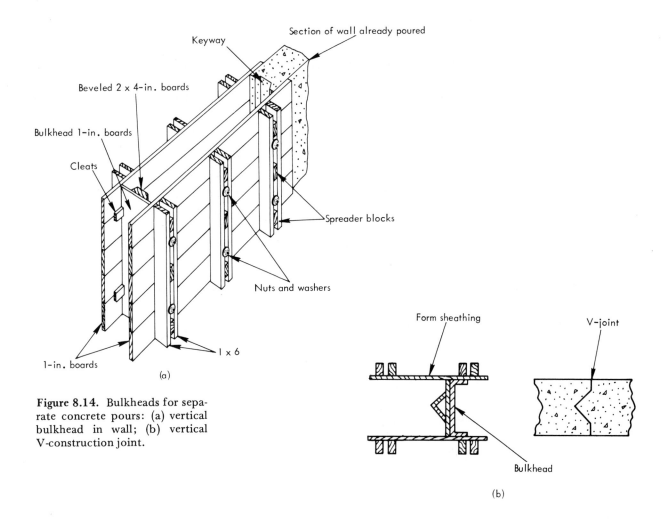

Figure 8.14. Bulkheads for separate concrete pours: (a) vertical bulkhead in wall; (b) vertical V-construction joint.

Figure 8.15. Anchor bolt details:
(a) hooked anchor bolt;
(b) anchor bolt with pipe sleeve;
(c) suspended anchor bolt.

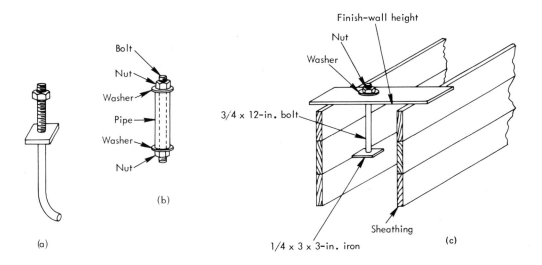

8.3.2   Footings

Footings for foundation walls are almost always of poured concrete. It is the simplest material to use and its plasticity lends itself to varied shapes, heights, and widths. The methods for pouring footings are the same as for walls, as described above. Figures 8.4 and 8.5 illustrate how footing forms are to be placed. As the footings bear the entire weight of the structure, it is always a wise precaution to place two No. 4 ($\frac{1}{2}$-in.) deformed reinforcing rods about 3 in. from the bottom. These rods will serve to bridge any spot weakness in the concrete of the footing or in the underlying earth.

For the builder's information, the following illustration will show the amount of weight that a footing and the earth beneath it will have to bear in the case of a normal $2\frac{1}{2}$-story wood-framed residence.

Load (in pounds per square foot) for normal live and dead loads:

Roof—pitched	35	(this is for asphalt roofing)
Attic floor	56	
Second floor	60	
First floor	60	
Interior partitions	12	
	223 lb/ft²	

If the house is 25 ft wide, each foot of the front and rear footings will be called upon to carry 223 × $12\frac{1}{2}$ ft (half the width), or 2787 lb. To this must be added the weight of the exterior wall, which may be approximated at 40 lb/ft², which for a 20-ft-high wall will amount to 800 lb. (a solid masonry wall can double this). The total weight is therefore 2787 + 800 = 3587 lb, which is well below the 2-ton/ft² bearing allowed for the No. 9 soil, as described in Table 2.2. The concrete footing itself can easily bear this weight. Obviously, a wider house, heavier materials, and a heavier live load (more people, heavy furnishings, filing cabinets, heavy equipment) will change these bearing loads very considerably.

8.3.3   Walls

The two most widely used materials for foundation walls are concrete and concrete masonry blocks. The usual code requirement is for 8- to 10-in. concrete or 8- to 12-in. concrete blocks, depending on the height and weight of the structure.

Concrete, Formwork, and Placement

Concrete walls require formwork, which must be tight and braced sufficiently to withstand the stress that it will be subjected to as the wet concrete is poured, and also the pressure of the wet concrete. Wet concrete weighs about 140 lb/cu³. A square foot of concrete in an 8-in. wall will weigh 140 × $\frac{8}{12}$ = 93 lb. The form of a 7-ft-high wall will have to support 93 × 7 = 651 lb per running foot. If the builder proposes to build the forms, Figure 8.16 shows that the required materials will be 2 x 4 studs on 10-in.

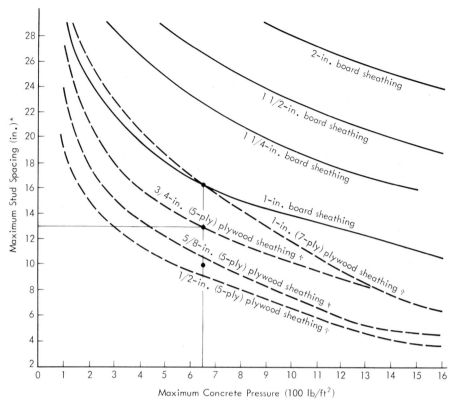

Figure 8.16.
Form-building guide.

Maximum Stud Spacing (in.)*

2-in. board sheathing

1 1/2-in. board sheathing

1 1/4-in. board sheathing

1-in. board sheathing

3, 4-in. (5-ply) plywood sheathing †

1-in. (7-ply) plywood sheathing †

5/8-in. (5-ply) plywood sheathing †

1/2-in. (5-ply) plywood sheathing †

Maximum Concrete Pressure (100 lb/ft²)

*Maximum allowable stud spacing = 32 in.
† Sanded face grain parallel to span.

Figure 8.17.
Bracing of wall forms.

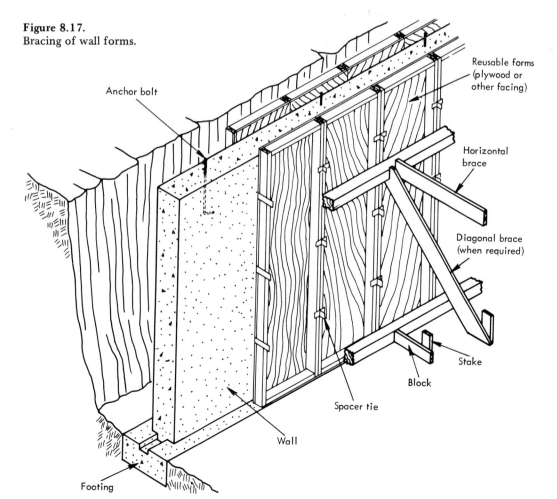

Anchor bolt

Reusable forms
(plywood or
other facing)

Horizontal
brace

Diagonal brace
(when required)

Stake

Block

Spacer tie

Wall

Footing

107

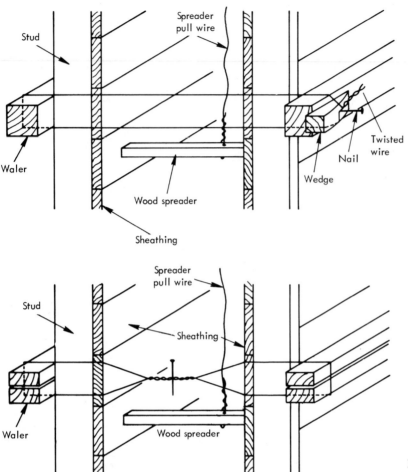

Figure 8.18. Detail showing spreaders and wire ties.

centers with ⅝-in. plywood, on 12-in. centers with ¾-in. plywood, or on 16-in. centers with 1-in. plywood; or board sheathing. If the builder builds even as few as five houses a year and has storage space, it would be much more economical to build unit forms which can be reused many times. It may even be more economical for the single-house builder to rent such forms. Such forms are built in panels that can be of any size—a popular one being 2 ft wide by 8 ft high. Well-built unit panels do not require as much bracing as do job-made forms. Figures 8.17 and 8.18 show methods of erecting and bracing wall forms.

In building forms for concrete walls, the builder must erect and place sturdy frames for door or window openings and should block out any necessary openings for utility lines, fuel lines, or other necessary openings. Cutting holes through concrete is expensive. The forms should be well oiled and the forms must be smooth on the concrete side.

Concrete Masonry

Foundation walls built of concrete masonry blocks are perfectly satisfactory for usual residential construction and may in some cases be less expensive than concrete and formwork. Concrete block (masonry) units

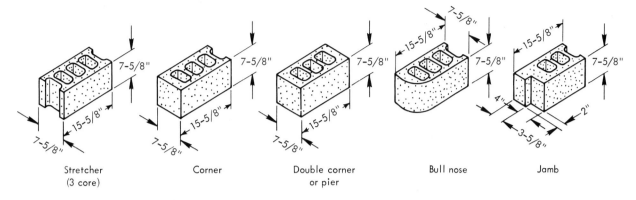

Figure 8.19. Standard concrete masonry blocks.

Stretcher (3 core)   Corner   Double corner or pier   Bull nose   Jamb

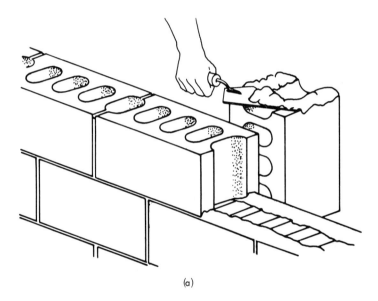

(a)

Figure 8.20. Laying of concrete block: (a) Vertical edges of blocks are "buttered" with mortar before laying; (b) method of holding block in placing it in position against block previously laid.

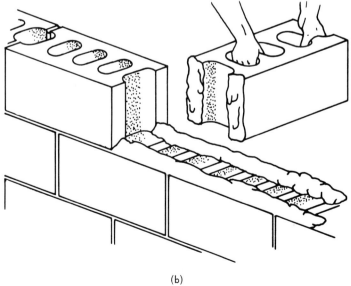

(b)

for foundations come in standard sizes, as shown in Figure 8.19. The blocks shown are for an 8-in. wall. The thickness of the block is 11⅝ for a 12-in. wall. Well-laid masonry walls can be made as water-resistant as concrete. They must be laid with full bed joints and full head joints, as shown in Figure 8.20 and the joints should be tooled smooth. For the nonprofessional who wishes to lay such a wall, it is a good idea to lay out the dry blocks on the footing to see how they can best be laid with the minimum of cutting. Table 8.4 shows how many full and half-blocks are required to produce certain lengths of wall. It is to be noted that these figures depend on the mason using a ⅜-in.-thick head joint. Table 8.5 shows how many courses have to be laid to produce a certain height. According to Table 8.5, it will take 11 courses of block (6'-8" + 8" = 7'-4") plus a 4-in. solid course on top, 7'-4" + 4" = 7'-8", to produce a 7'-4" height of wall above a 4-in. basement slab. If the builder wants only a 7'-0" clear height, the 4-in. top course can be omitted and, instead, the top course of blocks solidly filled with cement mortar.

The number of blocks to be used in a wall is shown in Table 8.6. It shows that if the builder uses the usual 7⅝ by 7⅝ by 15⅝-in. block for an 8-in. wall or the 11⅝-in.-thick block for a 12-in. wall, 112.5 blocks will be required for 100 ft² of wall. The heavy aggregate blocks that are required weigh 7950 lb. To use this as an example:

The foundation wall previously shown is 25 x 40 x 7 ft high.

This makes a perimeter of 130 ft or a total area of 910 ft².

Allowing 10% for waste and breakage, the builder would order block enough for 1000 ft² or 112.5 × 10 = 1125 blocks. The builder will use 8.5 ft³ of mortar per 100 blocks or a total of 8.5 × 11 = 93.5 ft³ of mortar.

TABLE 8.4. Nominal Length of Concrete Masonry Walls by Stretchers

Number of Stretchers	Units 15⅝ in. Long and Half Units 7⅝ in. Long with ⅜ in. Thick Head Joints	Number of Stretchers	Units 15⅝ in. Long and Half Units 7⅝ in. Long with ⅜ in. Thick Head Joints
1	1'-4"	8½	11'-4"
1½	2'-0"	9	12'-0"
2	2'-8"	9½	12'-8"
2½	3'-4"	10	13'-4"
3	4'-0"	10½	14'-0"
3½	4'-8"	11	14'-8"
4	5'-4"	11½	15'-4"
4½	6'-0"	12	16'-0"
5	6'-8"	12½	16'-8"
5½	7'-4"	13	17'-4"
6	8'-0"	13½	18'-0"
6½	8'-8"	14	18'-8"
7	9'-4"	14½	19'-4"
7½	10'-0"	15	20'-0"
8	10'-8"	20	26'-8"

TABLE 8.5. Nominal Height of Concrete Masonry Walls by Courses

Number of Courses	Nominal Height Units $7\frac{5}{8}$ in. High and $\frac{3}{8}$-in.-Thick Bed Joint	Number of Courses	Nominal Height Units $7\frac{5}{8}$ in. High and $\frac{3}{8}$-in.-Thick Bed Joint
1	8″	10	6′–8″
2	1′–4″	15	10′–0″
3	2′–0″	20	13′–4″
4	2′–8″	25	16′–8″
5	3′–4″	30	20′–0″
6	4′–0″	35	23′–4″
7	4′–8″	40	26′–8″
8	5′–4″	45	30′–0″
9	6′–0″	50	33′–4″

TABLE 8.6. Weights and Quantities of Materials for Concrete Masonry Walls[a]

Actual Unit Sizes (Width × Height × Length) (in.)	Nominal Wall Thickness (in.)	For 100 ft² of Wall		
		Number of Units	Average Weight of Finished Wall	
			Heavyweight Aggregate (lb)	Lightweight Aggregate (lb)
$7\frac{5}{8}$ × $7\frac{5}{8}$ × $15\frac{5}{8}$	8	112.5	5500	3600
$11\frac{5}{8}$ × $7\frac{5}{8}$ × $15\frac{5}{8}$	12	112.5	7950	4900

[a]Table based on $\frac{3}{8}$-in. mortar joints.

8.3.4   Foundations for Crawlways and Slabs on Ground

In areas where drainage is poor or the water table is high, a house with a crawlway is preferred to a full basement. In many of the warmer areas of the country, the great majority of houses are built on slabs on ground. The advantage of both these methods is, of course, the saving in cost by the omission of a basement with its excavation, foundation walls, its possible water problems, the avoidance of ledge rock, and the fact that a well-built and reinforced slab may avoid settlement problems. With the advent of packaged heating/air-conditioning units and compact laundry units, the basement is not as necessary as it once was. Storage facilities can be provided elsewhere.

Crawlways

When a crawlway is provided, it should be high enough for a person to "crawl" under the house for the purpose of inspecting or repairing heating, water, or electrical lines or for inspecting for termites. A good recommended height is 3′-6″ clear. The crawlway should be ventilated and the bottom of it be above groundwater level. It is recommended also that insulation be installed on the underside of the first-floor joists (see Chapter 13).

The foundation walls where crawlways are used are usually made of masonry blocks. The footings are made of concrete and the depth of footing depends on the frost line, which can range down to 4'-0" below grade in the northern states. Just enough excavation is required to bring the bottom down to 3'-6" below the underside of the first-floor joists. Columns to support first-floor girders and so on are built of concrete block on concrete footings.

Slabs on Ground

The footings for slabs on ground, like footings for crawlways, must go down below the frostline. Slabs on ground have been built for some time in all parts of the country, and because many were not properly built they have proven unsatisfactory and the source of many complaints. The floors have been cold, condensation has collected on the floor near cold or warm exterior walls, and in many cases of radiantly heated floors the entire under-floor heating coils have had to be abandoned because of leakage and the lag in response to changes in outside temperature.

New techniques in slab construction have changed all this and a properly built slab floor is now perfectly satisfactory. Figures 8.21 to 8.24 show some of the methods used in present-day construction of slab floors. The basic requirements for a satisfactory slab floor are:

Finish-floor level to be high enough above the existing grade so that fill can be provided to drain water away from the house. Although 8-in. clearance above finish grade is shown in the illustrations, the author recommends a clearance of 12 in. With only 8 in. of clear foundation wall, it is very easy to miss carpenter ants or termites, and 12 in. provides that extra margin of safety against flooding in heavy rainstorms or seepage from snow banks.

There must be a vapor barrier between the slab and the underlying fill to prevent moisture from seeping into the concrete. The barrier should be strong enough to resist puncturing when the concrete is poured over it. Possibilities are 45-lb asphalt roofing paper or at least 6-mil polyethylene sheeting lapped at least 6 in. and mopped with hot asphalt or troweled mastic.

Insulation should be provided around the perimeter walls. It should be rigid, nonabsorptive, and waterproof. The table shown in Table 8.7 shows the recommended resistance (R) factor for various exterior temperatures. The R factor and other insulating facts will be explained in Chapter 13.

The slab should be reinforced with 6 × 6 No. 10 wire mesh and should be at least 4 in. thick. In any case where there is a question about the quality of the underlying earth, or just for better construction, a 6-in. slab is recommended and it should be done in one pour.

If a finish floor is to be applied directly to the top surface of the slab, it should be carefully leveled and screeded and then should be steel-troweled after it has been floated.

In warmer parts of the country where frost penetration is not a problem, the footing and slab can be poured as a monolithic structure. However, the footings must be placed on solid undisturbed ground.

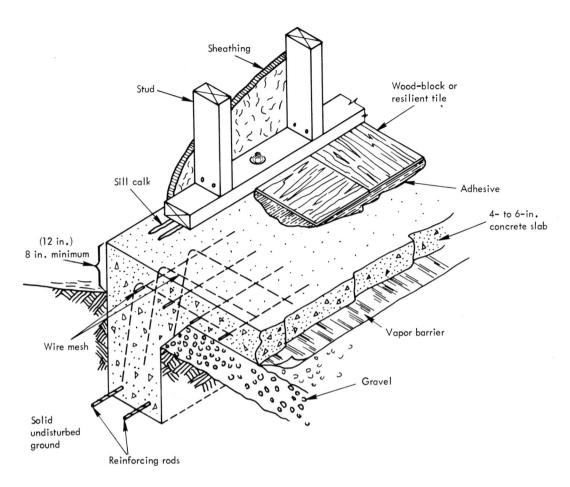

Sheathing

Stud

Wood-block or
resilient tile

Sill calk

Adhesive

(12 in.)
8 in. minimum

4- to 6-in.
concrete slab

Wire mesh

Vapor barrier

Solid
undisturbed
ground

Gravel

Reinforcing rods

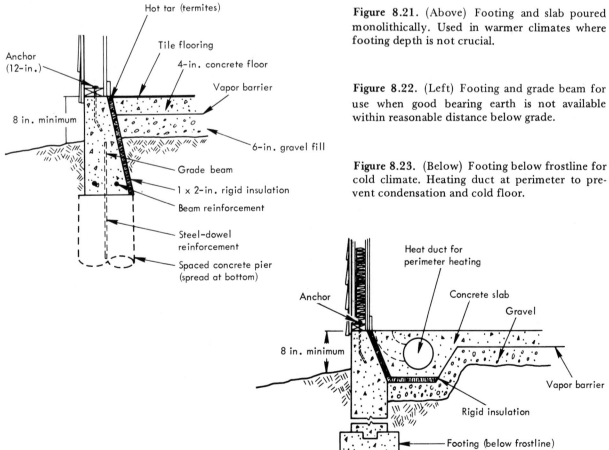

Hot tar (termites)

Tile flooring

Anchor
(12-in.)

4-in. concrete floor

Vapor barrier

8 in. minimum

6-in. gravel fill

Grade beam

1 x 2-in. rigid insulation

Beam reinforcement

Steel-dowel
reinforcement

Spaced concrete pier
(spread at bottom)

Figure 8.21. (Above) Footing and slab poured monolithically. Used in warmer climates where footing depth is not crucial.

Figure 8.22. (Left) Footing and grade beam for use when good bearing earth is not available within reasonable distance below grade.

Figure 8.23. (Below) Footing below frostline for cold climate. Heating duct at perimeter to prevent condensation and cold floor.

Heat duct for
perimeter heating

Anchor

Concrete slab

Gravel

8 in. minimum

Vapor barrier

Rigid insulation

Footing (below frostline)

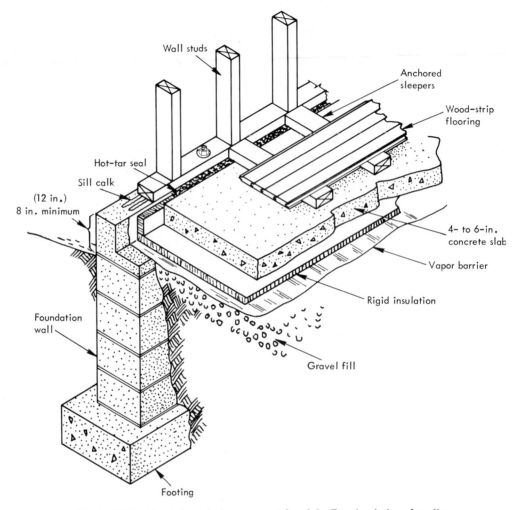

Figure 8.24. Block foundation support for slab. Footing below frostline. Wood flooring on sleepers is superior construction.

TABLE 8.7. Resistance Values Used in Determining Minimum Amount of Edge Insulation for Concrete Floors Slabs on Ground for Various Design Temperatures

Low Temperatures (°F)	Depth Insulation Extends Below Grade (ft)	Resistance (R) Factor	
		No Floor Heating	Floor Heating
−20	2	3.0	4.0
−10	1½	2.5	3.5
0	1	2.0	3.0
+10	1	2.0	3.0
+20	1	2.0	3.0

8.3.5 Foundation Reinforcement

The commonly poured 8- or 10-in. concrete foundation wall for a single-family residence does not normally require reinforcement. If there are openings in the wall, two No. 4 (½-in.) reinforcing rods laid 1½ in. above the opening as the wall is being poured will be satisfactory unless the opening is wider than normal door or window size (Figure 8.25). In such cases the builder can use precast reinforced-concrete lintels or steel lintels as a support for the wall above. Such lintels are also used in masonry walls. Reinforcing rods should also be used in concrete footings where any soft spot in the underlying earth is suspected, and they should be used in cases where two separate parts of a structure are poured at different times and at different elevations (main house and garage, etc.). The rods serve to form a connecting bond between them. In such cases the use of three rods placed horizontally is recommended (Figure 8.26). Steel reinforcement for concrete structures, for retaining walls, and for other heavy load-bearing concrete structures should be designed by a structural engineer.

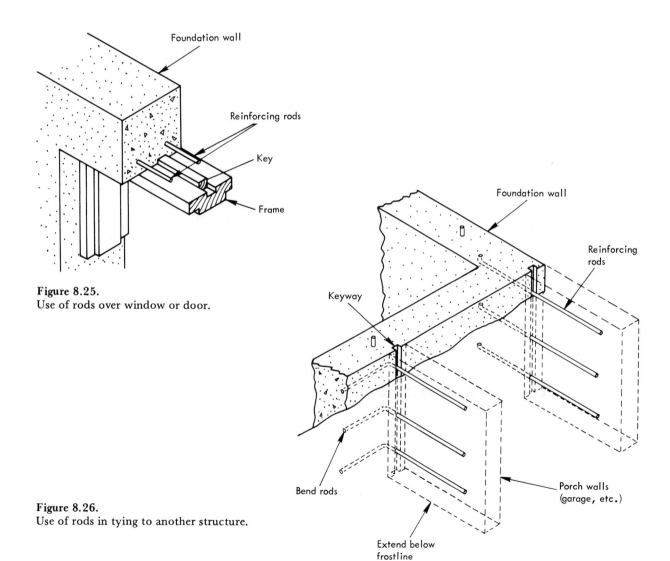

Figure 8.25.
Use of rods over window or door.

Figure 8.26.
Use of rods in tying to another structure.

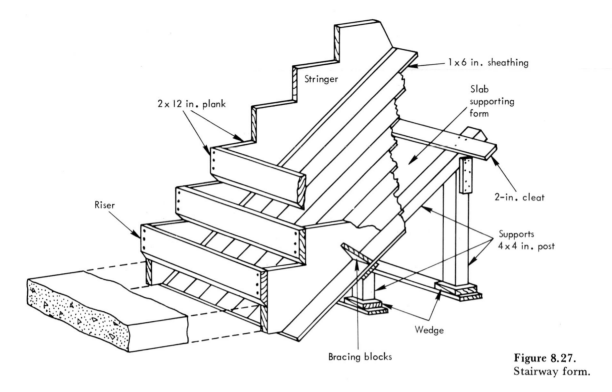

Figure 8.27.
Stairway form.

8.3.6   Miscellaneous Concrete

In residential construction concrete is of course used in many structural parts other than footings or foundation walls. It is used for self-supporting floor slabs, for basement floors, for stairs, for retaining walls, and for many other purposes that require the plasticity of concrete. Figure 8.27 shows the method of constructing a form for a concrete stairway. A free-standing stair such as this is usually reinforced by bars laid parallel to and above the "slab supporting form." The basement floor should be at least 4 in. thick (6 in. is much better) and should be reinforced with 6 X 6 No. 10 wire mesh. The concrete should be poured as soon as the foundation walls are complete. The bottom of the basement slab should rest on top of the footings and should, where necessary, have a perimeter of a waterproofing substance.

8.3.7   Cost Estimating for Foundations

The estimator has first to establish what materials are to be used for footings and foundation walls. He or she must also decide whether deep footings and concrete, or masonry columns or piles for poor soil conditions will be used. The estimator must choose whether ready-mix concrete will be purchased or concrete mixed at the site. Table 8.8 is a sample of an estimating breakdown.

The first step in completing the estimate is to establish quantities. The perimeter of the foundation wall multiplied by the height and thickness will give cubic feet, and this divided by 27 will give cubic yards. For example:

Perimeter $\times$ height $\times$ thickness in feet $\div$ 27 = cubic yards

Perimeter $\times$ height $\div$ 100 $\times$ concrete block per 100 ft^2 = number of blocks

Perimeter = running feet of foundation drain

Perimeter $\times$ 1 ft wide $\times$ 1 ft deep $\div$ 27
$$= \text{cubic yards of crushed stone for drain.}$$

Total length of rebars + lbs/ft
$$= \text{total pounds} \times \text{cost of material and labor/lb}$$

The second step is to establish the labor unit cost.

The third step is to obtain current prices for material. As a (possibly foolish) caution, the estimator should be careful to check the arithmetic for amount and decimal points!

TABLE 8.8. Summary of Foundation Costs

	Unit Labor	Unit Material	Equipment	Total Unit Cost	Total Units	Total Cost
General footings						
Forms						
Reinforcement						
Concrete						
Miscellaneous footings, chimneys, piers						
Forms						
Reinforcement						
Concrete						
Deep footings						
Forms						
Reinforcement						
Concrete						
Concrete masonry						
Grade beams						
Foundation walls						
Forms						
Poured concrete						
Concrete masonry						
Anchors						
Footing drains						
Waterproofing						
Basement slab						
Slab on grade						
Gravel or stone bed						
Forms						
Vapor barrier						
Insulation						
Reinforcement rod or mesh						
Poured concrete						
Finishing						
Footings and foundations (miscellaneous)						
Stairways						
Areaways						
Accessory structures						

CHAPTER NINE

structural framing

9.1   THE PRELIMINARY PREPARATION

The preparation for the structural framing should begin before any construction has started. The builder who has prepared an estimate and submitted a bid for a house or who is building it on speculation will certainly have prepared lumber lists, hardware lists, and millwork lists. The builder-owner of the single house must also do this so that first, he or she will be prepared to order material as it is required; second, so that a wholesale price can be obtained; and third, so that he or she can produce a total cost figure in order to obtain a construction loan and a mortgage.

The preceding points are true for the entire construction process, but advance preparation is particularly important for the structural frame, which should be completed and closed in in as continuous an operation as possible.

9.2   FRAMING MATERIALS

All framing lumber is manufactured from softwoods such as Douglas fir, southern pine, eastern and western spruce, and eastern and western hemlock. Structural or framing lumber is graded by its strength; its freedom from imperfections such as knots, checks, shakes, or slope of grain that would diminish its strength; and by its moisture content. General construction and utility lumber is also known as common lumber. Most reliable construction lumber is grade-stamped and the builder should become familiar with the meaning of these stamps. Lumber is graded by such trade groups as the Western Wood Products Association, Southern Pine Inspection Bureau,

TABLE 9.1. Sizes of Dimension Lumber

Nominal	Unseasoned	Dry	Area (Dry) (in.^{2})
2 × 2	$1\frac{9}{16} \times 1\frac{9}{16}$	$1\frac{1}{2} \times 1\frac{1}{2}$	2.25
2 × 3	$1\frac{9}{16} \times 2\frac{9}{16}$	$1\frac{1}{2} \times 2\frac{1}{2}$	3.75
2 × 4	$1\frac{9}{16} \times 3\frac{9}{16}$	$1\frac{1}{2} \times 3\frac{1}{2}$	5.25
2 × 6	$1\frac{9}{16} \times 5\frac{5}{8}$	$1\frac{1}{2} \times 5\frac{1}{2}$	8.25
2 × 8	$1\frac{9}{16} \times 7\frac{1}{2}$	$1\frac{1}{2} \times 7\frac{1}{4}$	10.87
2 × 10	$1\frac{9}{16} \times 9\frac{1}{2}$	$1\frac{1}{2} \times 9\frac{1}{4}$	13.87
2 × 12	$1\frac{9}{16} \times 11\frac{1}{2}$	$1\frac{1}{2} \times 11\frac{1}{4}$	16.87
4 × 4	$3\frac{9}{16} \times 3\frac{9}{16}$	$3\frac{1}{2} \times 3\frac{1}{2}$	12.25
4 × 6	$3\frac{9}{16} \times 5\frac{5}{8}$	$3\frac{1}{2} \times 5\frac{1}{2}$	19.25
4 × 8	$3\frac{9}{16} \times 7\frac{1}{2}$	$3\frac{1}{2} \times 7\frac{1}{4}$	25.37

National Lumber Manufacturers Association, and others. The local wholesale lumber dealer should have brochures that explain the meaning of these stamps.

The selection and proper use of construction hardware is just as important as the selection of lumber. No structure is better than its connections and a weak link anywhere, even in the most insignificant place, can cause quite serious consequences.

9.2.1   Lumber: Sizes, Grades, and Quality

All construction framing lumber is manufactured in standard sizes both as to width, thickness, and length. Wood for trim, flooring, and other miscellaneous uses will be discussed in following chapters. Table 9.1 lists the sizes of the so-called "dimension lumber" that is most commonly used. It will be noted that the nominal size (the name by which the lumber is called such as 2 x 4, 2 x 6, etc.) is not the size that the builder will get from the lumber dealer. The table gives the nominal size, the size when unseasoned, and the final size, which is what the builder will get. Table 9.2 gives the sizes for boards that are used among other places for decking and for roofing and flooring in post-and-beam framing. The builder must be fully aware of these actual sizes as well as the modular lengths of lumber, which do not vary and range from 6 ft through 2-ft additions to 24 ft. The sizes of the lumber control the sizes of bracing, blocking, rough framing, girders, and so on.

TABLE 9.2. Sizes of Boards

Nominal	Unseasoned	Dry
1 × 4	$\frac{3}{4} \times 3\frac{9}{16}$	$1\frac{1}{16} \times 3\frac{1}{2}$
1 × 6	$\frac{3}{4} \times 5\frac{5}{8}$	$1\frac{1}{16} \times 5\frac{1}{2}$
1 × 8	$\frac{3}{4} \times 7\frac{1}{2}$	$1\frac{1}{16} \times 7\frac{1}{4}$
1 × 10	$\frac{3}{4} \times 9\frac{1}{2}$	$1\frac{1}{16} \times 9\frac{1}{4}$
1 × 12	$\frac{3}{4} \times 11\frac{1}{2}$	$1\frac{1}{16} \times 11\frac{1}{4}$

Grades and Quality

Common lumber is graded by the use to which it will be put.

The best grade (standard) is suitable for use without waste. It contains only tight knots and can be used where it will sustain the maximum stress.

Construction grade is of the same general quality as the best grade. This is the lumber that is used most often in structural framing. It allows well-spaced knots and limited checking but none at the end.

Economy grade is used for footing forms and rough flooring, and as utility lumber for guard rails and other purposes where it must be cut into nonrecoverable short lengths.

The moisture content of lumber is very important. The builder is advised to use only lumber that is stamped kiln-dried, or "S-dry," which indicates a maximum of 19% moisture with an average of 15%, or "MC 15" which indicates a maximum moisture content of 15% and averages 10 to 12%.

9.2.2   Recommended Uses; Structural Strength

The structural strength of framing lumber depends on the kind of wood and the quality. The woods used for framing lumber that have the highest ratings for resistance to shear and bending and the highest modulus of elasticity* are Douglas fir and southern pine, closely followed by western hemlock, Norway pine, and larch. Any of these woods which are available at a competitive price can be used for framing in construction quality. The moisture content is especially important in members such as rafters, joists, and girders, where shrinkage across their width can cause settlement and cracking.

9.2.3   Use of Modular Sizes

As all structural lumber comes in certain fixed sizes, the designer and builder should dimension and build the structure so as to avoid as much waste and labor as possible by using as much of the wood as they can. For instance, a 14-ft joist will rest on 2 x 4 plates (or a ribbon) on the exterior and 2 x 4 plates over an interior bearing partition. The full bearing length will be 8 in. at most. Therefore, the clear span or room size can be 14 ft less 8 in. or 13 ft 4 in. Studs come in 8-ft lengths. If they rest on a plate (platform framing) and support a double plate on top to support ceiling joists, the entire height from rough floor to underside of joist will be (using actual sizes) 3 x 1½ = 4½ in. plus 8 ft for the stud, or a total of 8 ft 4½ in. This is too high. If the builder wishes to sheath the exterior with plywood, which comes in 8-ft heights, the studs must be cut to fit this dimension. The sheathing can be run from the bottom of the sole to the top of the top plates and the stud cut to 7'-7½'' (Figure 9.1). The only waste is in the short pieces of stud. The joists and the plywood can be nailed in place without cutting. The same adherence to modular length can be used for girders, rafters, sole plates, sills, and any other structural members. To repeat—it will save material and time.

*The modulus of elasticity (E) is the ratio of the unit stress to the corresponding unit deformation. It is the ability of the material to resist deformation. The higher the modulus, the better the structural quality.

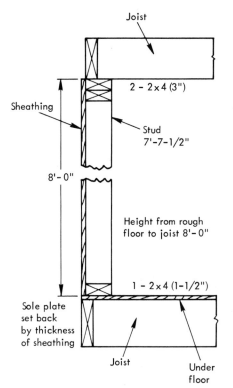

Joist

2 – 2 x 4 (3")

Sheathing

Stud
7'-7-1/2"

8'- 0"

Height from rough
floor to joist 8'- 0"

1 – 2 x 4 (1-1/2")

Sole plate
set back
by thickness
of sheathing

Joist

Under
floor

Figure 9.1.
Use of modular sizes.

Figure 9.2.
Nail sizes.

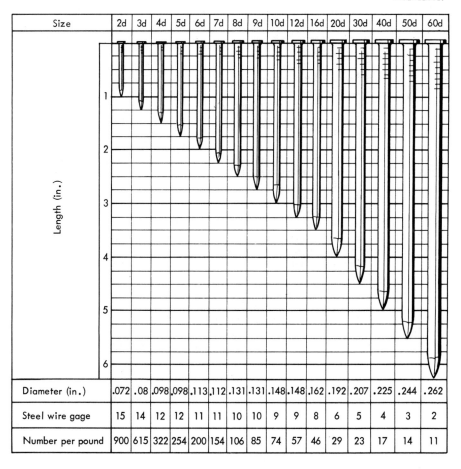

Size	2d	3d	4d	5d	6d	7d	8d	9d	10d	12d	16d	20d	30d	40d	50d	60d
Diameter (in.)	.072	.08	.098	.098	.113	.112	.131	.131	.148	.148	.162	.192	.207	.225	.244	.262
Steel wire gage	15	14	12	12	11	11	10	10	9	9	8	6	5	4	3	2
Number per pound	900	615	322	254	200	154	106	85	74	57	46	29	23	17	14	11

Length (in.)

TABLE 9.3. Nailing Schedule for a Well-Constructed Wood-Frame House

Joining	Nailing Method	Nails		
		Number	Size	Placement
Header to joist	End-nail	3	16d	
Joist to sill or girder	Toenail	2	10d or	
		3	8d	
Header and stringer joist to sill	Toenail		10d	16 in. on center
Bridging to joist	Toenail each end	2	8d	
Ledger strip to beam, 2 in. thick		3	16d	At each joist
Subfloor, boards:				
1 by 6 in. and smaller		2	8d	To each joist
1 by 8 in.		3	8d	To each joist
Subfloor, plywood:				
At edges			8d	6 in. on center
At intermediate joists			8d	8 in. on center
Subfloor (2 by 6 in., T&G) to joist or girder	Blind-nail (casing) and face-nail	2	16d	
Soleplate to stud, horizontal assembly	End-nail	2	16d	At each stud
Top plate to stud	End-nail	2	16d	
Stud to soleplate	Toenail	4	8d	
Soleplate to joist or blocking	Face-nail		16d	16 in. on center
Doubled studs	Face-nail, stagger		10d	16 in. on center
End stud of intersecting wall to exterior wall stud	Face-nail		16d	16 in. on center
Upper top plate to lower top plate	Face-nail		16d	16 in. on center
Upper top plate, laps and intersections	Face-nail	2	16d	
Continuous header, two pieces, each edge			12d	12 in. on center
Ceiling joist to top wall plates	Toenail	3	8d	
Ceiling joist laps at partition	Face-nail	4	16d	
Rafter to top plate	Toenail	2	8d	
Rafter to ceiling joist	Face-nail	5	10d	
Rafter to valley or hip rafter	Toenail	3	10d	
Ridge board to rafter	End-nail	3	10d	
Rafter to rafter through ridge board	Toenail	4	8d	
	Edge-nail	1	10d	
Collar beam to rafter:				
2-in. member	Face-nail	2	12d	
1-in. member	Face-nail	3	8d	
1-in. diagonal let-in brace to each stud and plate				
(4 nails at top)		2	8d	
Built-up corner studs:				
Studs to blocking	Face-nail	2	10d	Each side
Intersecting stud to corner studs	Face-nail		16d	12 in. on center
Built-up girders and beams, three or more members	Face-nail		20d	32 in. on center, each side
Wall sheathing:				
1 by 8 in. or less, horizontal	Face-nail	2	8d	At each stud
1 by 6 in. or greater, diagonal	Face-nail	3	8d	At each stud
Wall sheathing, vertically applied plywood:				
$\frac{3}{8}$ in. and less thick	Face-nail		6d	6 in. edge
$\frac{1}{2}$ in. and over thick	Face-nail		8d	12 in. intermediate
Wall sheathing, vertically applied fiberboard:				
$\frac{1}{2}$ in. thick	Face-nail		$1\frac{1}{2}$-in. roofing nail	3-in. edge and
$\frac{25}{32}$ in. thick	Face-nail		$1\frac{3}{4}$-in. roofing nail	6-in. intermediate
Roof sheathing, boards, 4-, 6-, 8-in. width	Face-nail	2	8d	At each rafter
Roof sheathing, plywood:				
$\frac{3}{8}$ in. and less thick	Face-nail		6d	6-in. edge and
$\frac{1}{2}$ in. and over thick	Face-nail		8d	12-in. intermediate

TABLE 9.4. Wood-Screw Sizes

Size Numbers

Length (in.)	0	1	2	3	4	5	6	7	8	9	10	11	12	13	14	15	16	17	18	20	22	24	26	28	30
1/4		X	X	X																					
3/8	X	X	X	X																					
1/2		X	X	X	X																				
5/8		X	X	X	X																				
3/4			X	X	X	X	X	X	X	X	X	X	X	X	X	X	X								
7/8			X	X	X	X	X	X	X	X	X	X	X	X	X	X	X								
1				X	X	X	X	X	X	X	X	X	X	X	X	X	X	X	X						
1 1/4					X	X	X	X	X	X	X	X	X	X	X	X	X	X	X	X	X	X			
1 1/2					X	X	X	X	X	X	X	X	X	X	X	X	X	X	X	X	X	X			
1 3/4						X	X	X	X	X	X	X	X	X	X	X	X	X	X	X	X	X			
2						X	X	X	X	X	X	X	X	X	X	X	X	X	X	X	X	X			
2 1/4						X	X	X	X	X	X	X	X	X	X	X	X	X	X	X	X	X			
2 1/2						X	X	X	X	X	X	X	X	X	X	X	X	X	X	X	X	X	X		
2 3/4							X	X	X	X	X	X	X	X	X	X	X	X	X	X	X	X	X		
3									X	X	X	X	X	X	X	X	X	X	X	X	X	X	X		
3 1/2									X	X	X	X	X	X	X	X	X	X	X	X	X	X	X		
4									X	X	X	X	X		X		X	X	X	X	X	X	X	X	X
4 1/2													X		X		X		X	X	X	X	X	X	X
5													X		X		X	X	X	X	X	X	X	X	X
6															X		X		X	X	X	X	X	X	X

9.2.4 Framing Hardware

Research and good practice have shown that the nails and screws, stud bolts and lag bolts, and joist fasteners and other connecting devices to be used must be of the correct size and used in the proper amount to ensure that the building will be structurally sound.

Figure 9.2 is a graphic picture of nail sizes. It gives their length, diameter, and wire gauge size and the number of nails per pound. Table 9.3 shows a nailing schedule. Many codes contain nailing schedules, but this one is a consensus of several codes. Table 9.4 is a summary of screw sizes and dimensions, and Table 9.5 gives the standard sizes of lag screws. Figure 9.3 shows the various types of carriage bolts and Table 9.6 their sizes.

TABLE 9.5. Lag-Bolt Sizes

Lengths (in.)	Diameters (in.)				
	$\frac{1}{4}$	$\frac{3}{8}, \frac{7}{16}, \frac{1}{2}$	$\frac{5}{8}, \frac{3}{4}$	$\frac{7}{8}, 1$
1	X	X		
$1\frac{1}{2}$	X	X	X	
2, $2\frac{1}{2}$, 3, $3\frac{1}{2}$, etc., $7\frac{1}{2}$, 8 to 10	X	X	X	X
11 to 12		X	X	X
13 to 16			X	X

TABLE 9.6. Carriage-bolt Sizes

Lengths (in.)	Diameters (in.)		
	$\frac{3}{16}, \frac{1}{4}, \frac{5}{16}, \frac{3}{8}$	$\frac{7}{16}, \frac{1}{2}$	$\frac{9}{16}, \frac{5}{8}, \frac{3}{4}$
$\frac{3}{4}$	X		
1	X	X	
$1\frac{1}{4}$	X	X	X
$1\frac{1}{2}$, 2, $2\frac{1}{2}$, etc., $9\frac{1}{2}$, 10 to 20	X	X	X X

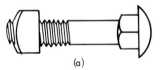

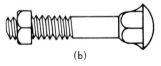

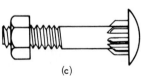

Figure 9.3.
Types of carriage bolts:
(a) square or common;
(b) finned neck;
(c) ribbed neck.

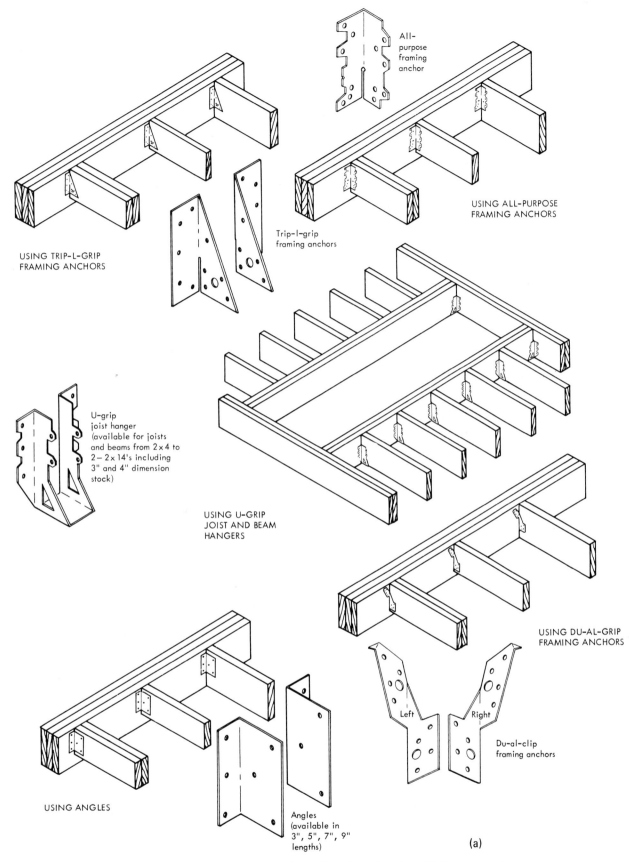

All-purpose framing anchor

USING ALL-PURPOSE FRAMING ANCHORS

USING TRIP-L-GRIP FRAMING ANCHORS

Trip-l-grip framing anchors

U-grip joist hanger (available for joists and beams from 2 x 4 to 2 – 2 x 14's including 3" and 4" dimension stock)

USING U-GRIP JOIST AND BEAM HANGERS

USING DU-AL-GRIP FRAMING ANCHORS

Left   Right

Du-al-clip framing anchors

USING ANGLES

Angles (available in 3", 5", 7", 9" lengths)

(a)

Figure 9.4. Fastening devices: (a) for floor-level framing: (b) for roof anchorage; (c) and (d) miscellaneous. (Courtesy TECO, 5530 Wisconsin Avenue, Washington, D.C. 20015).

125

All anchors shown may be used either with conventional rafters or with trusses.

Anchor may often be placed on inside of plate, if desired, if proper bearing and nailing surfaces are available.

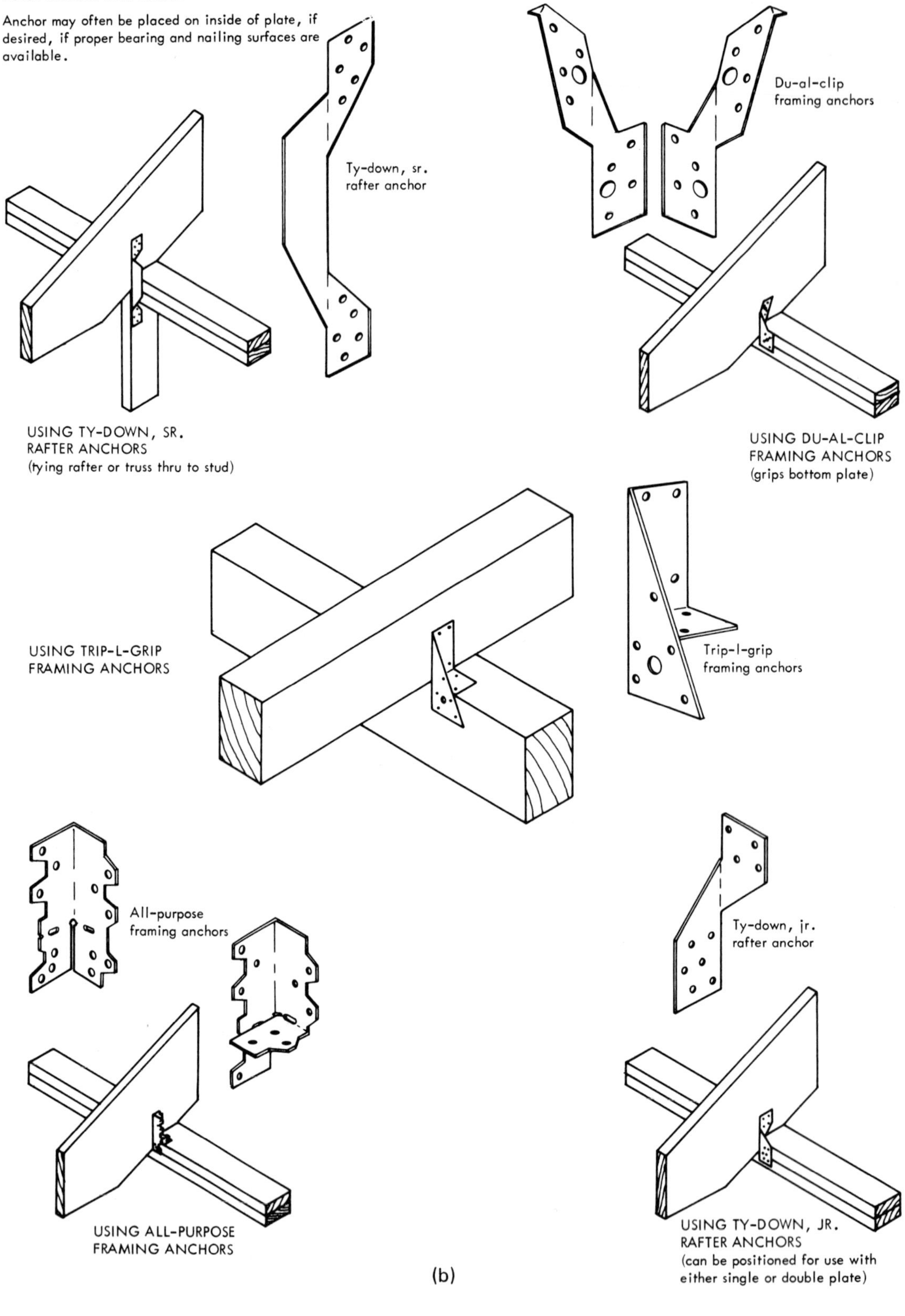

Ty-down, sr. rafter anchor

Du-al-clip framing anchors

USING TY-DOWN, SR. RAFTER ANCHORS
(tying rafter or truss thru to stud)

USING DU-AL-CLIP FRAMING ANCHORS
(grips bottom plate)

USING TRIP-L-GRIP FRAMING ANCHORS

Trip-l-grip framing anchors

All-purpose framing anchors

Ty-down, jr. rafter anchor

USING ALL-PURPOSE FRAMING ANCHORS

USING TY-DOWN, JR. RAFTER ANCHORS
(can be positioned for use with either single or double plate)

(b)

HUD's MPS   Paragraph 606-4.7 page 6-6-9 says:
"h. A continuous header sized according to the span of the largest opening
may be used in lieu of a double top plate. Heading members shall be
connected at corners and at intersecting bearing partitions with sheet
metal corner ties, lag screws or other suitable means."

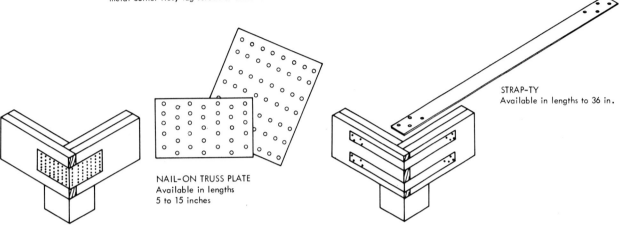

NAIL-ON TRUSS PLATE
Available in lengths
5 to 15 inches

STRAP-TY
Available in lengths to 36 in.

and the Manual of Acceptable Practices to HUD MPS
Article 606-4 Post and Beam connection says:
"Where sheathing does not anchor posts to plates, metal anchors
should be provided as shown. Roof beams must also be anchored
to posts. Walls must be braced for resistance racking."

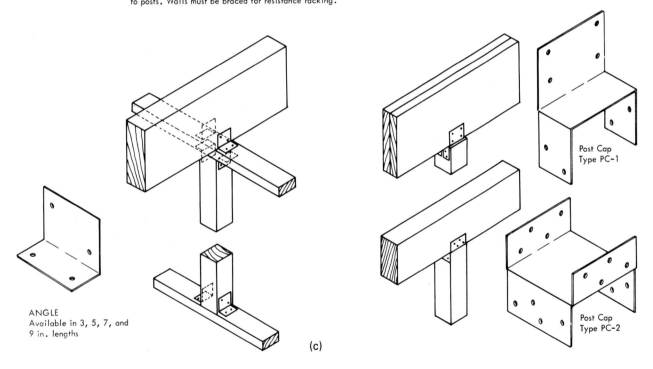

ANGLE
Available in 3, 5, 7, and
9 in. lengths

(c)

Post Cap
Type PC-1

Post Cap
Type PC-2

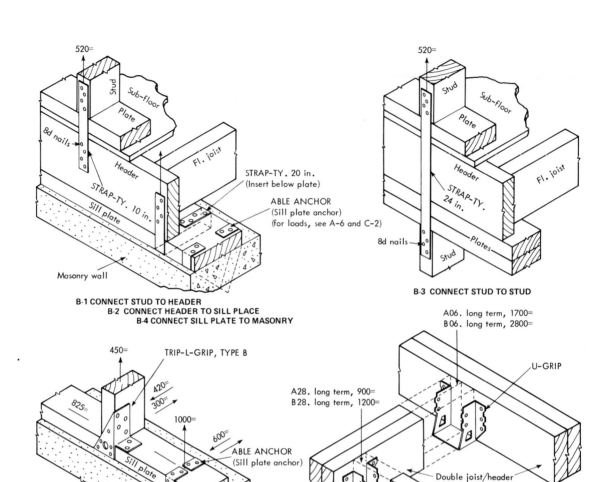

520=

Stud

Sub-floor

Plate

8d nails

Fl. joist

STRAP-TY. 10 in.

Header

STRAP-TY. 20 in. (Insert below plate)

ABLE ANCHOR (Sill plate anchor) (for loads, see A-6 and C-2)

Sill plate

Masonry wall

B-1 CONNECT STUD TO HEADER
B-2 CONNECT HEADER TO SILL PLACE
B-4 CONNECT SILL PLATE TO MASONRY

520=

Stud

Sub-floor

Plate

Header

Fl. joist

STRAP-TY. 24 in.

8d nails

Plates

Stud

B-3 CONNECT STUD TO STUD

450=

TRIP-L-GRIP, TYPE B

420=

300=

825=

1000=

600=

Sill plate

ABLE ANCHOR (Sill plate anchor)

Masonry (found, wall or slab floor)

C-1 CONNECT STUD TO SILL PLATE
C-2 CONNECT SILL PLATE TO MASONRY

A06. long term, 1700=
B06. long term, 2800=

U-GRIP

A28. long term, 900=
B28. long term, 1200=

Double joist/header /1 piece beam

Joist

U-GRIP

D-1 JOIST SUPPORT PLUS VERTICAL AND LATERAL TIE

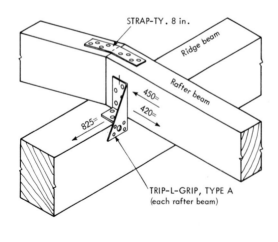

STRAP-TY. 8 in.

Ridge beam

Rafter beam

450=

420=

825=

TRIP-L-GRIP, TYPE A (each rafter beam)

E-1 CONNECT RAFTER BEAMS TO RIDGE BEAM

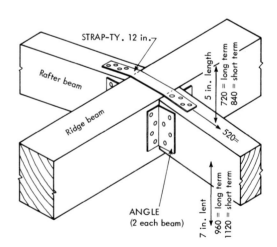

STRAP-TY. 12 in.

Rafter beam

5 in. length
720 = long term
840 = short term

Ridge beam

520=

ANGLE (2 each beam)

7 in. lent
960 = long term
1120 = short term

E-2 CONNECT RAFTER BEAM TO RIDGE BEAM

(d)

128

Several general rules for the use of nails and screws are:

Fewer nails that are properly sized and driven will hold better than a great many nails that are driven close together.

Nails driven across the grain hold better than those driven with the grain.

Screws hold better than nails and can draw the connected pieces closer together. They cost more and take more labor.

Before a screw is used, a hole of the diameter of the screw should be drilled in the first piece and a hole smaller than the screw diameter and not the full depth of the screw should be driven in the second piece, to prevent splitting the wood.

Lag bolts are sturdier than screws and can be used for critical connections.

Carriage bolts are used for positive wood-to-wood connections.

The hardware mentioned above is used essentially for wood-to-wood connections. There is another type of framing hardware which is used to connect wood to wood by means of metal devices which are shaped so that they can be fastened to the framing members and hold them together by the strength of the device itself. Various figures in this chapter show such devices, which are used especially in post-and-beam framing. Figure 9.4 shows some of the standard shapes. They are available at any materials dealer.

9.3 THE ELEMENTS OF STRUCTURAL FRAMING

The framing of the great majority of residential construction is done with wood. Steel I-beams may be used instead of wood girders for long spans; steel or precast concrete lintels are used over openings in masonry construction or over wide openings in wood construction; lally columns are used as columns to support girders, and expanded steel joists are sometimes used in large residential construction.

9.3.1 Wall Framing

The most commonly used types of wall framing are the western or platform frame and the balloon frame. Each has its advantages and disadvantages.

Platform Frame

Because the platform frame enables the builder to erect the first floor immediately, it has the great advantage of immediately providing a working surface for the erection of the wall frame. The exterior wall framing and the interior partitions can be built in sections on the floor and then pushed up into place. This is a saving in labor cost that the builder should not overlook. In an attempt at assembly-line construction, some builders will construct an entire exterior wall assembly on the floor. This can include exterior sheathing and even window frames and sash. The danger of this method is the possible difficulty of bringing the wall to plumb after it is assembled, and extra care must be taken to assure this as the wall is being constructed.

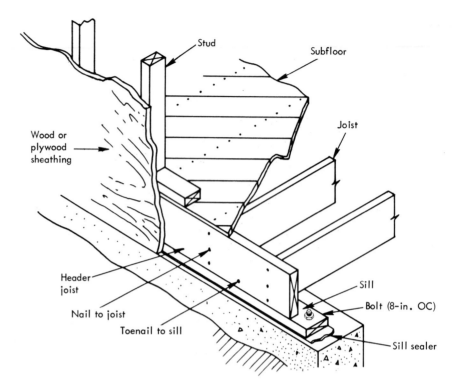

Figure 9.5.
Platform framing at sill.

Figure 9.6. Platform framing:
Assembled wall sections.

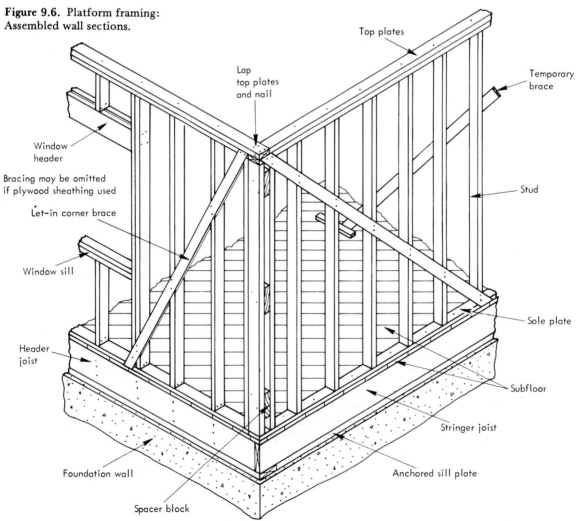

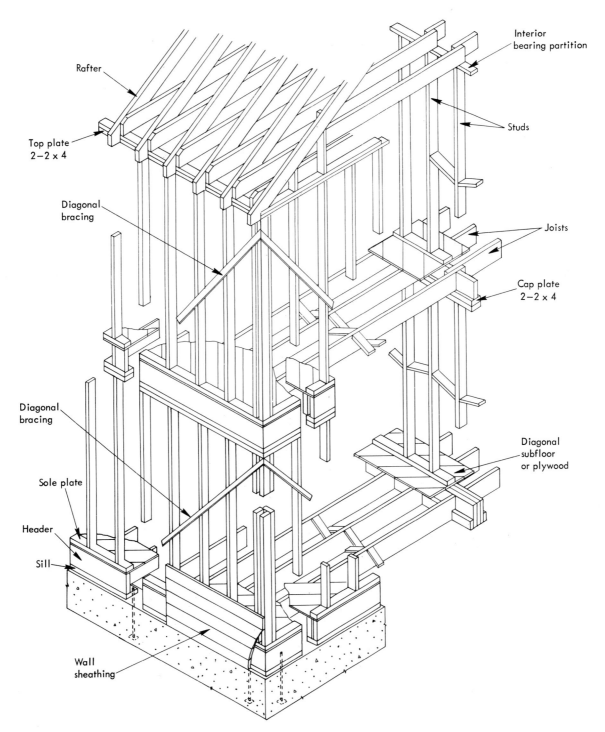

Rafter

Top plate
2—2 x 4

Diagonal
bracing

Diagonal
bracing

Sole plate

Header

Sill

Wall
sheathing

Interior
bearing partition

Studs

Joists

Cap plate
2—2 x 4

Diagonal
subfloor
or plywood

Figure 9.7.
Platform framing. House assembly.

131

The disadvantage of platform framing is that framing lumber shrinks as it dries. As most of the shrinkage occurs horizontally in the sills, header joists, and top and bottom plates, this can cause some future cracking. If the builder is careful about the moisture content of the lumber, this will not be too serious. In a single-story house the total vertical dimension of the lumber whose shrinkage is important is as follows in nominal dimensions: sill, 2 in.; header, 8 or 10 in.; plates, three at 2 in.; for a total of $2 + 10 + 6 = 18$ in. The joists will shrink at the same rate as the header and the interior partition and exterior wall plates will also shrink at the same rate, which will help to alleviate the problem. Figure 9.5 shows a detailed view of platform framing at the sill. Figure 9.6 shows wall sections that have been assembled on the floor and pushed up into place. Figure 9.7 shows an entire wall assembly and Figure 9.8 shows how the end or gable wall is framed.

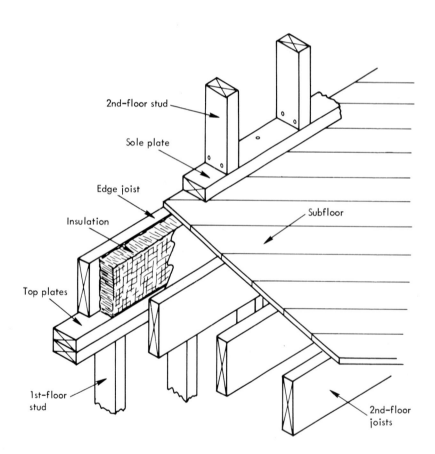

2nd-floor stud

Sole plate

Edge joist

Insulation

Subfloor

Top plates

1st-floor stud

2nd-floor joists

Figure 9.8.
Framing of end wall: Platform.

Balloon Frame

In balloon framing the exterior wall studs rest on the sill and the joists are nailed directly to them. The floor framing is done first; then some underflooring must be put down so that the carpenter can use it as a platform as the walls are framed. Also in this framing the builder must fit a fire stop between each stud to prevent an updraft in the space between the studs, which would otherwise be completely open to the basement.

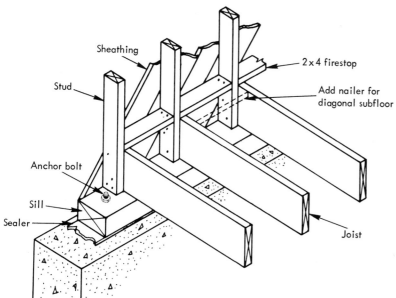

Sheathing

Stud

Anchor bolt

Sill

Sealer

2 x 4 firestop

Add nailer for
diagonal subfloor

Joist

Figure 9.9. Balloon framing at sill.

Figure 9.10. Balloon framing wall detail.

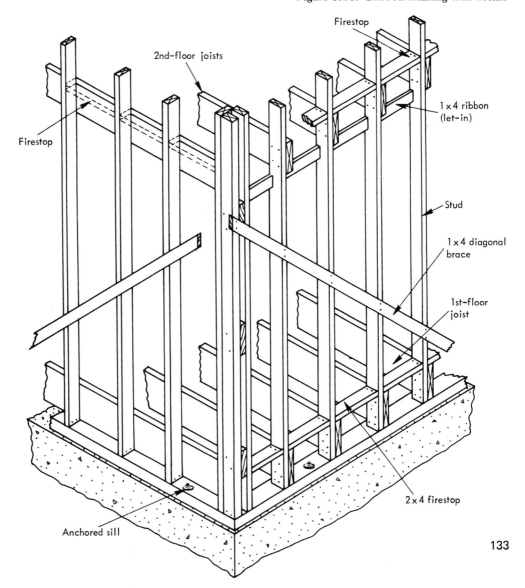

Firestop

2nd-floor joists

Firestop

1 x 4 ribbon
(let-in)

Stud

1 x 4 diagonal
brace

1st-floor
joist

2 x 4 firestop

Anchored sill

133

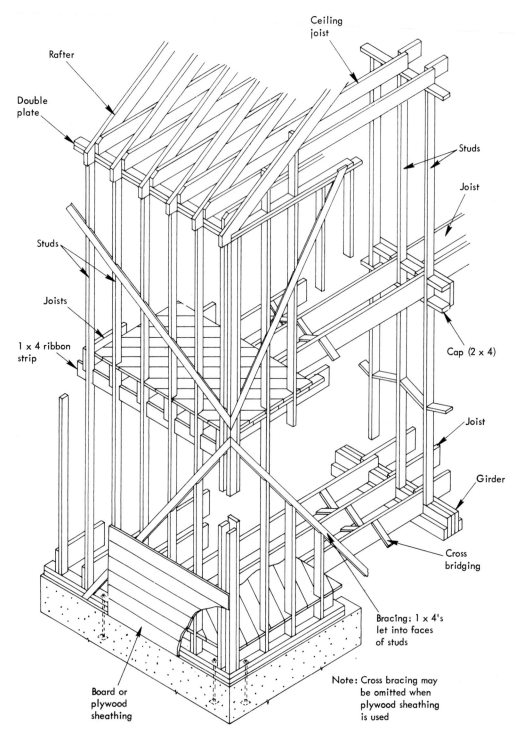

Figure 9.11.
Balloon framing assembly.

Rafter

Ceiling joist

Double plate

Studs

Joist

Studs

Joists

Cap (2 x 4)

1 x 4 ribbon strip

Joist

Girder

Cross bridging

Bracing: 1 x 4's let into faces of studs

Board or plywood sheathing

Note: Cross bracing may be omitted when plywood sheathing is used

134

The second floor joists in this framing rest on a ribbon strip instead of a platform. Figure 9.9 shows the details of balloon framing at the sill. Figure 9.10 shows a first-floor assembly which, unlike platform framing, has to be built in place. An entire balloon frame assembly is shown in Figure 9.11. The framing of the end wall or gable wall is shown in Figure 9.12.

The main advantage of balloon framing is that there is no horizontal shrinkage and that the vertical shrinkage in the supporting studs is approximately the same for every stud.

Subflooring. In platform framing the subfloor must be installed before any plates can be put down and, as noted previously, the floor acts as a working platform. In balloon framing the subfloor should be installed as soon as all floor joists are down and nailed in place. Subflooring can be of square edge or tongue-and-groove boards ¾ in. thick of economy-grade lumber. It can be laid horizontally or diagonally. To avoid warping, the boards should not be more than 8 in. wide.

It has become fairly common practice to use plywood for subflooring. It comes in large sheets and is therefore labor saving. In Structural 1 and 2 and C-C grades, it has structural rigidity so that it serves as a bracing member of the floor installation. The tables shown in Section 9.5 give the proper spacing between joists for various grades and thicknesses of plywood. As an addition to these tables, it is to be noted that Douglas fir (coast type) and southern pine will allow a ½-in. thickness for 16-in. spacing, ⅝ in. for 20-in. spacing and ¾ in. for 24-in. spacing. The lesser structural grades, such as western hemlock, western white pine, and ponderosa pine require ⅝ in., ¾ in., and ⅞ in. for the same spans.

Plywood subflooring comes with one side dressed so that floor tile or carpeting can be applied to it directly.

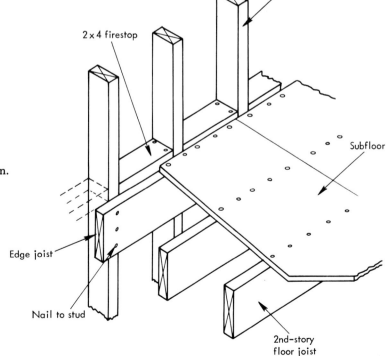

Figure 9.12. Framing of end wall: Balloon.

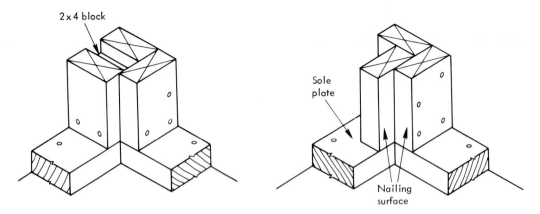

Figure 9.13. Examples of corner post assemblies.

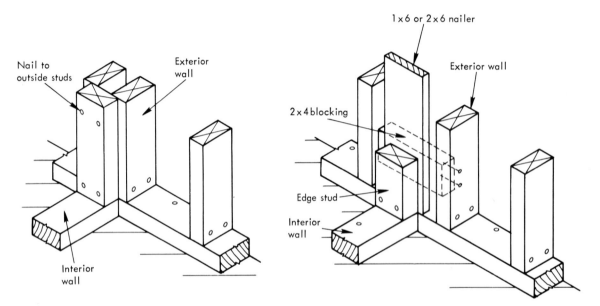

Figure 9.14. Interior partition butting and going through an exterior wall.

Figure 9.15. Types of wall bridging.

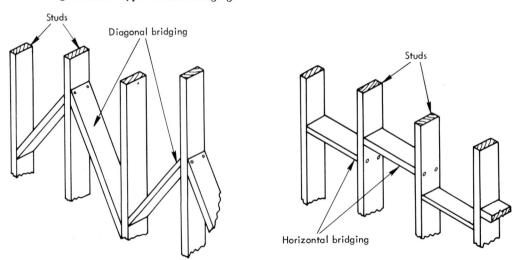

Corner Framing and Bracing

The structural strength of the wall frame depends to a large extent on its corner assemblies and its bracing. With the use of wall sheathing made of materials with no structural strength (unlike wood sheathing), it is necessary to corner-brace most houses with a 1 x 4 ribbon let into the studs. Bracing or bridging between the studs is also important. Figure 9.13 shows two corner assemblies. Figure 9.14 shows approved framing where an interior partition abuts an exterior wall or goes through it, and Figure 9.15 shows two methods of bracing or bridging between exterior studs.

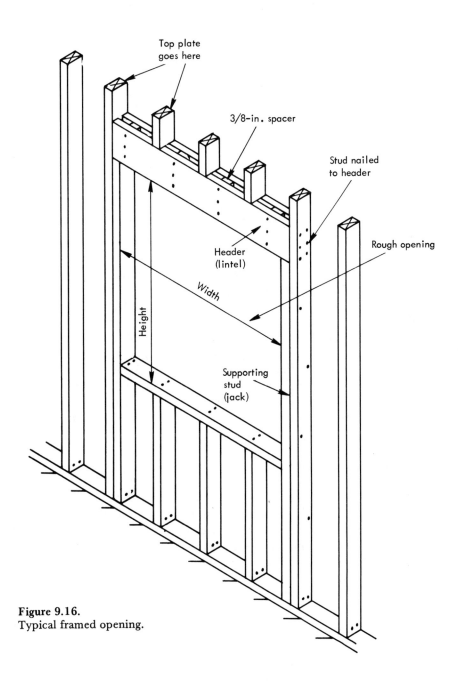

Figure 9.16.
Typical framed opening.

Window and Door Framing

If the builder plans a platform-type frame, the exterior wall openings can be assembled before the wall is "pushed up" or these openings can be framed along with the balloon framing. In any case, all openings in the exterior wall require extra framing to provide the structural strength that is lost by the opening. Figure 9.16 shows that a window (or door opening) requires double studs at each side—as shown, the outside studs go up to the top plate and the inside studs serve as jacks to support the header or lintel. The lintel must be strong enough to support the upper floor and roof over the width of the opening. The local building code will usually give the size of the lintel that is required for openings of various widths. Following is a list of recommended lintel sizes:

Maximum Span (ft)	Header or Lintel Size (in.)
3½	2–2 × 6
5	2–2 × 8
6½	2–2 × 10
8	2–2 × 12

Light structural steel is recommended for any opening wider than 8 feet.

The builder must also make openings in the rough frame that are large enough to allow room for the window and door frames and trim to be set in place, and to be properly leveled and properly wedged. Figures 9.17 and 9.18 show the standard method of framing for door and window openings. It is suggested that the rough-framed openings for double-hung windows should be 6 in. wider than the sash width and 10 in. higher than the combined height of the upper and lower sash. Rough framing for casement windows should be 11¼ in. wider than the combined *glass* width and 6⅜ in. higher than the *glass* height. Rough framing for doors should be 2½ in. wider than the width of door and 3 in. higher than the height of the door.

How to Plumb a Wall

After a wall is assembled with its studs and top and bottom plates and before it is braced or sheathed, it must be plumbed. This must be done before the walls, which meet at right angles, are permanently fastened together. The best and quickest way to plumb a wall is by the use of a surveyor's transit, but such an instrument is not always available and requires some training in its use. The use of a plumb bob and rule is an excellent way and can be done very simply (Figure 9.19):

Fasten the plumb-bob cord to a rule and have it long enough to fall to the bottom of the corner post.

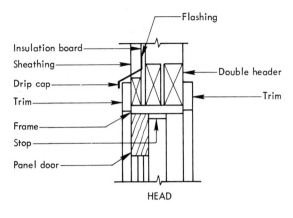

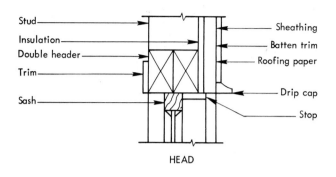

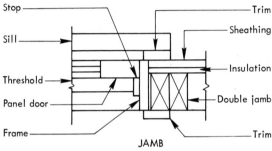

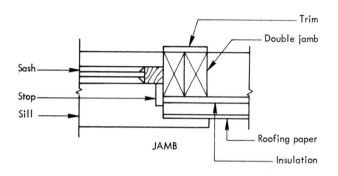

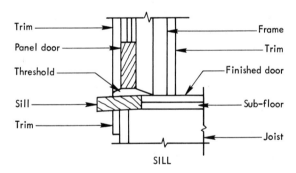

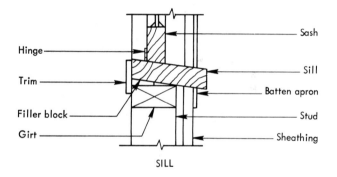

Figure 9.17.
Typical exterior door framing.

Figure 9.18.
Typical exterior window framing.

Lay the rule along the top plate and have the plumb-bob cord fall 2 in. from the edge of the post.

Lay another rule alongside the top edge of the bottom plate and project it 2 in. from the outer edge of the post. When the wall is plumb, the cord will be just touching the end of the rule.

Brace the wall with temporary braces until the finish sheathing or bracing is applied. The wall may also require inside bracing until the walls are joined at the corners. In balloon framing the plumb bob will go from the top edge of the ribbon to the top of the sill.

After the wall is plumbed, the builder can proceed with the permanent bracing or sheathing which will be illustrated in Chapter 10.

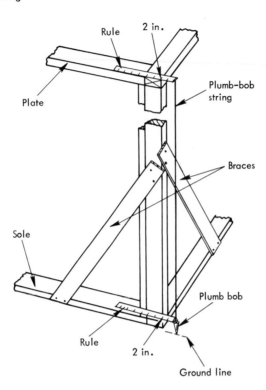

Figure 9.19.
Plumbing and bracing a wall.

9.3.2   Floor Framing

As has been previously mentioned, when platform framing is used, the entire first floor can be framed and rough-covered before the erection of the exterior walls is started. In balloon framing the floor is framed with the wall. In wood construction the floor joists are supported by girders in the interior of the house and by wood sills supported by foundation walls on the exterior perimeter or by pockets in the foundation wall. The various types of wooden girder supports for joists are shown in Figure 9.20. In Figure 9.20a and b, joists rest either entirely on ledger strips or are notched over the girder and are additionally supported by ledger strips. In these cases, especially in the first one, the joist must be thoroughly toe-nailed to the girder and the ledger must be carefully fastened. In Figure 9.20c, the joists rest on top of the girder. If there is a basement, the girder will project into the head room by its full width. If the foundation walls are full 8 ft high, the loss of headroom will not be important. This method is excellent over crawlways and is good anywhere when headroom is not important.

When a particularly long clear span is required, it is sometimes more economical to use a steel girder supported by wood or steel columns. Figure 9.21 shows a ledger bolted or on blocks resting on the flange. Either of these methods can also be used when the floor support consists of expanded steel joists. Girders are supported at the perimeter walls by the sills or by pockets in the concrete or masonry foundation walls, as shown in Figure 9.22. The choice depends on how the builder wishes to frame the joists to the girder.

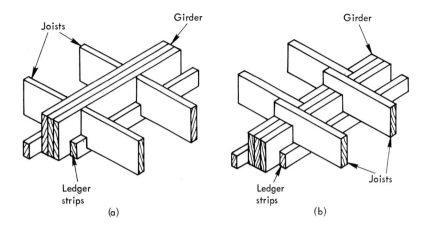

Figure 9.20. Use of wood girder.

(a)

(b)

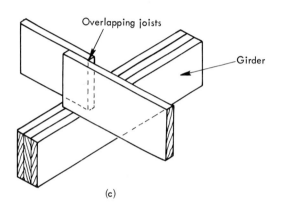

(c)

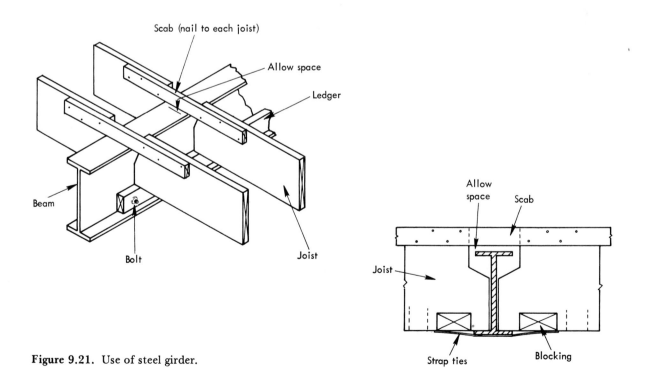

Figure 9.21. Use of steel girder.

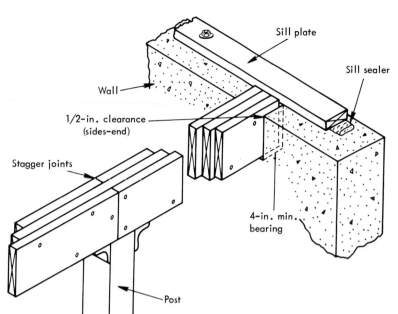

Figure 9.22. Girder supported by pocket in foundation wall. With 8'-0" foundation wall there is sufficient basement headroom.

In balloon framing of floors for upper stories, the joists are supported at the exterior walls by ribbons that are cut into the studs and by a single or double top plate over interior bearing partitions. In platform framing, the joists are supported by double top plates over the exterior wall studs and are nailed into a header beam. The interior bearing partitions are the same for both framing methods (Figure 9.23).

Framing of Floor Openings

Floor openings for stairwells, chimneys, ventilating ducts, or any other large construction that pierces a floor are framed by headers and trimmers. The size of the header and trimmer beams depends on the size of the opening and its location. In most cases it is recommended that steel beam hangers be used in addition to toenailing to provide positive support. Figure 9.24 shows various methods of constructing such framing. In Figure 9.24a, the opening is against an exterior wall and the trimmer runs parallel to the joists. In Figure 9.24b, the trimmer runs parallel with the joists and rests on the exterior wall and an interior bearing partition. Figure 9.24c and d show larger openings, which require both headers and double trimmers.

To keep the floor joists rigidly in place and at the correct distance from each other, the builder must install bridging between them. Three types of bridging are in general use. Figure 9.25 shows these types.

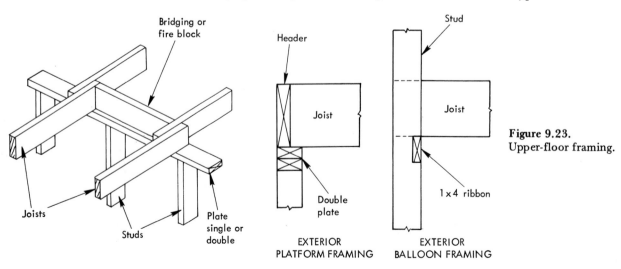

Figure 9.23. Upper-floor framing.

JOISTS OVER BEARING PARTITION

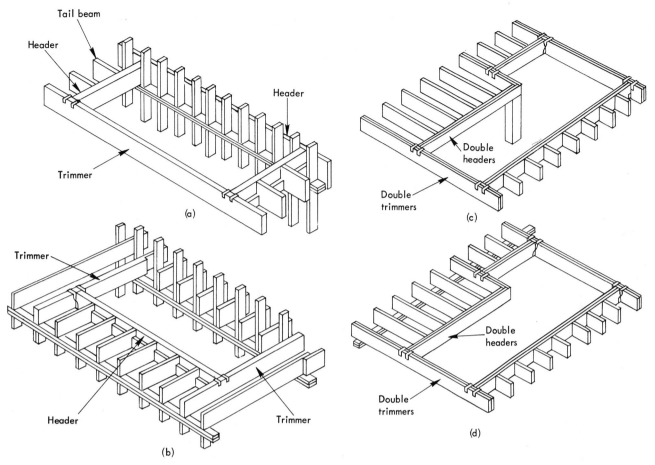

Figure 9.24. Methods of framing floor openings.

Figure 9.25. Types of bridging between joists.

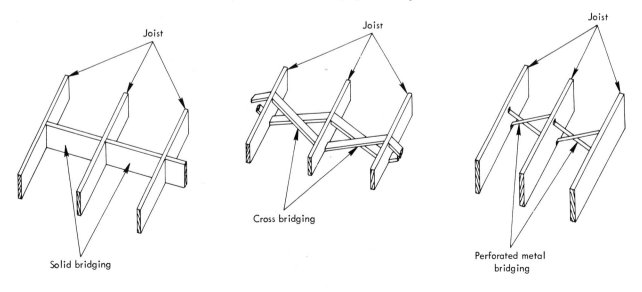

143

Floor Framing Summary: Step-by-Step Procedure

The various methods of the framing and the supporting of a floor have been explained above. This section will put them all together in a step-by-step progression.

Place sills and fasten to foundation wall by anchor bolts.

Erect girders to bear on sills or exterior wall pockets and on interior columns of masonry or steel.

Lay joists in place and nail joist cross bracing at one end only.

Place double joists under partitions that run parallel to them.

Nail header to joists (in platform framing) or nail exterior studs to joists and nail 2 x 4 firestops between studs and to tops of joists (in balloon framing).

Brace studs in balloon framing.

Toenail platform headers to sill.

Nail cross bracing at other end to make floor rigid.

Frame floor openings.

Lay subfloor of plywood or of diagonal boards.

Figure 9.26 shows a platform floor assembly.

Figure 9.26. Platform framed floor assembly.

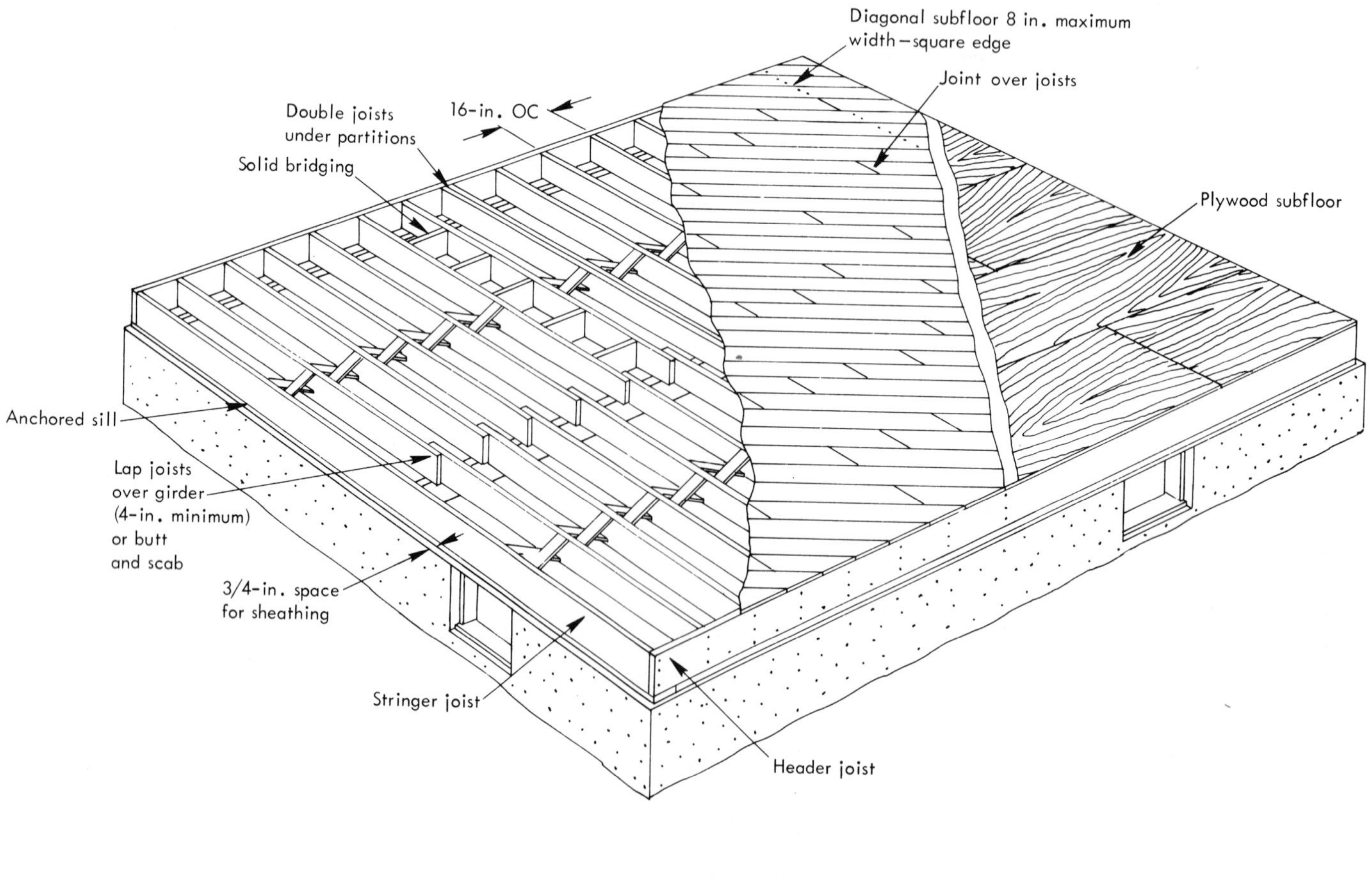

9.3.3 Partition Framing

Partitions serve as room dividers and as load bearers. In conventional construction the partitions are built of 2 x 4s which are on 16-in. centers. The usual partition can be framed and built on the floor and then raised into place. The partition is built in the same manner as the exterior wall, with a single bottom plate and two top plates.

If roof trusses are used in a single-story house, no load-bearing partitions need be used and the partitions can be fastened to the bottom chord of the trusses. In bearing partitions, openings for doors are framed with double studs, one being a supporting stud with a lintel above to bear the weight. Openings in nonbearing partitions can be framed with no supporting lintel. Figure 9.27 shows a framed opening in a bearing partition. Lintel sizes are the same as those listed in Section 9.3.1.

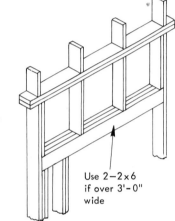

Figure 9.27.
Doorway framing in bearing partition.

Use 2−2 x 6
if over 3'−0"
wide

9.3.4 Ceiling and Roof Framing

When the exterior wall studs and the interior partition studs are erected in place and are braced and capped with double plates or connected by ribbon beams, the next step is the erection of the ceiling joists. These joists may support a second floor or an attic floor. To support a second floor in balloon framing the joists are supported by ribbon instead of on double plates as in platform framing. The ceiling framing over the space directly under the roof is the same for all framing—joists supported by bearing partitions on the interior and by double plates on the exterior.

The ceiling framing serves a double purpose. It supports the finish ceiling of the room below and the floor of the room above. In the case where the roof rafters are parallel to the ceiling joists, the joists should be firmly nailed together over the bearing partition and also be firmly nailed to the rafters so that the entire ceiling and roof structure can act as a truss to withstand the outward thrust of the rafters. The joists should also be toenailed to the top plates of the exterior wall and the interior partitions (Figures 9.28 to 9.30). In cases where the joists run at right angles to the rafters it is necessary to use collar beams and cross-ties to resist the outward thrust of the rafters.

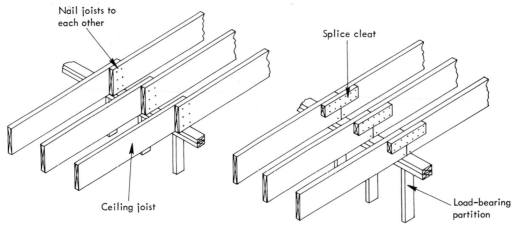

Nail joists to each other

Splice cleat

Ceiling joist

Load-bearing partition

Figure 9.28. Methods of securing joists.

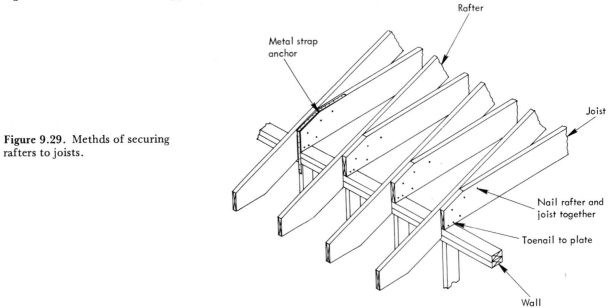

Figure 9.29. Methds of securing rafters to joists.

Rafter

Metal strap anchor

Joist

Nail rafter and joist together

Toenail to plate

Wall

Figure 9.30. Roof assembly.

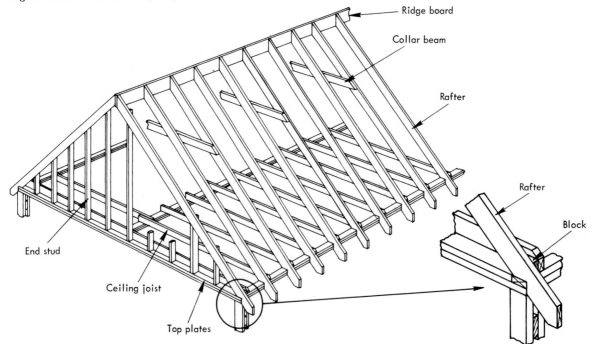

Ridge board

Collar beam

Rafter

Rafter

Block

End stud

Ceiling joist

Top plates

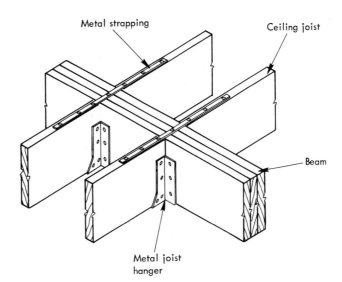

Figure 9.31. Girder and joist connection for a flush ceiling.

Ceiling framing over a wide opening can be done in two ways. If there is no objection to the look of a beam projecting from the ceiling, it can be done by a girder across the opening. The girder is supported at either end by a 4 x 4 or 6 x 6 or steel pipe column, and in turn it supports the ceiling joists at the same level as the bearing partition. If a completely flush ceiling between two spaces is required, it is best to use a flush ceiling system with all joists framing into a girder by means of metal joist hangers (Figure 9.31).

Figure 9.32 (a).
Method of framing a dormer.

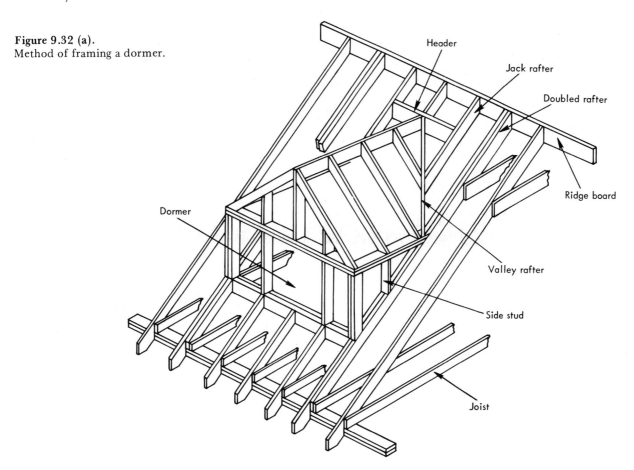

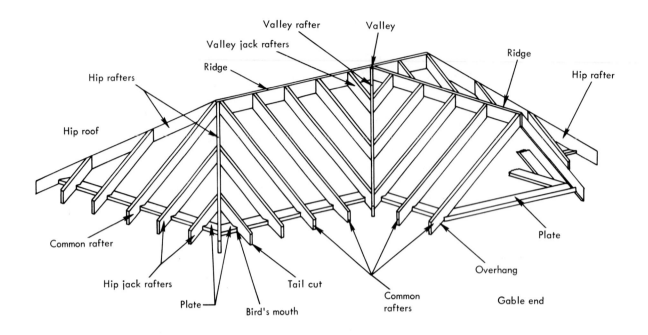

Figure 9.32 (b). Framing hip, valley, and gable.

Figure 9.33. Example of overhang for flat or low-pitched roof.

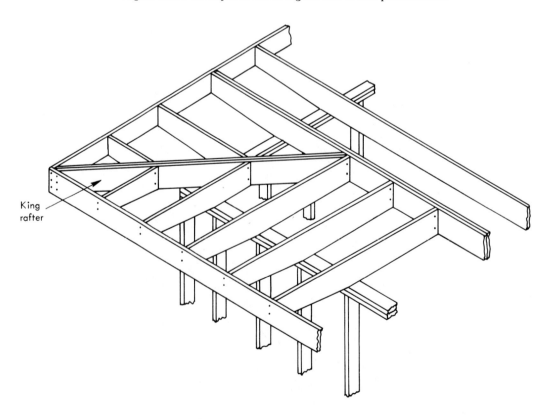

148

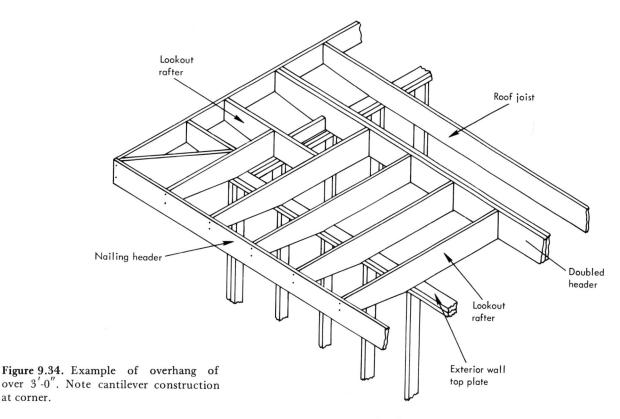

Figure 9.34. Example of overhang of over 3'-0". Note cantilever construction at corner.

The most common way of framing a roof is by means of rafters resting on exterior walls. This is fairly simple construction until the builder has to frame a dormer or join two roofs at right angles or frame a hip roof. Figure 9.32 shows how this is done. Figures 9.33 and 9.34 show examples of flat roof overhangs.

How to Determine Length of Rafters

The size and the spacing of rafters depends on the unsupported length of the rafters and the live load they are called upon to carry. Table 2.3 includes a list of allowable lengths for a roof truss. It is also necessary for the builder to determine rafter lengths for the bill of materials. Figure 9.35 shows a simple method for determining this.

Figure 9.35. Use of right-angle triangle to determine rafter length.

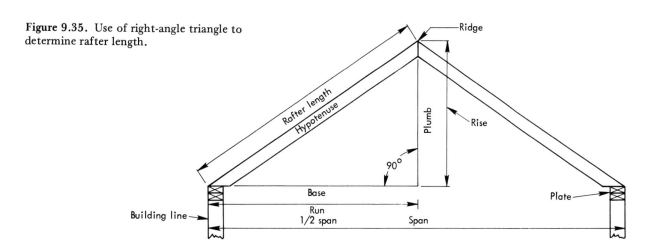

After the slope of the roof is determined, a simple right-angle triangle to scale can be ruled off. The bottom line is equal to one-half the span of the roof, and the vertical line at a right angle to the bottom is equal to the rise of the roof. The rafter length is the hypotenuse of the right-angle triangle. The formula is familiar to anyone who has had geometry. The length of the hypotenuse of a right-angle triangle is equal to the square root of the sum of the squares of the sides. Example: ½ span or run is 13 ft and rise is 6 ft 6 in. Length of rafter is square root of the total of 13 × 13 = 169 plus 6.5 × 6.5 = 42 for a total of 211. Taking the square root, we obtain 14.5 = 14 ft 6 in. (from a table of square roots, or use a computer).

Roof Trusses

Many builders now use yard-built prefabricated roof trusses or they can fabricate their own trusses on the site. The advantage of such a truss is that it can be built on the ground in an assembly-line way. The truss also uses smaller sizes of lumber. For instance, a truss built of 2 x 4s can span up to a 28-ft clear span. Of course, the trusses once made have to be hoisted into place, but this can be done with at most a 1-day use of a small crane. The builder must balance the labor and material cost of the conventional use of rafters with the cost of the trusses in place. Figure 9.36 shows how such a

Figure 9.36. Prefabricated roof truss. When 2 × 4s are used, intermediate gusset G may not be more than 10′–11″ from the end of the span.

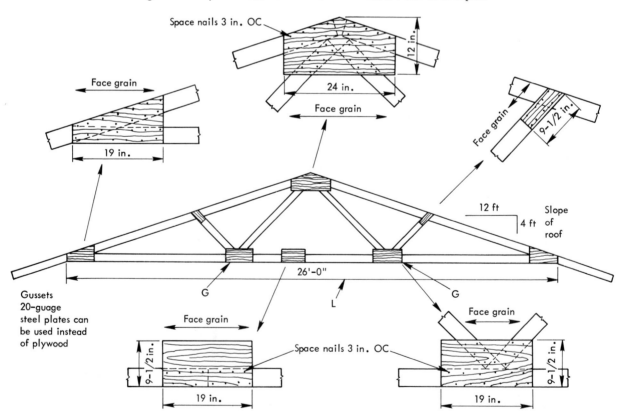

truss is built. The gusset plates are of ½-in. exterior-grade plywood and the lumber must be of good-quality pine or fir with a moisture content of not over 15%. The gussets should be both nailed and glued to both sides of the truss. It is suggested that moisture-resistant glues be used in areas of high humidity. Instead of plywood, the builder can use 20-gauge steel plates with nail holes drilled in them at 3-in. centers.

Slope of Roof (ft)	L for 2 × 4	L for 2 × 6
2 in 12	25	32
3 in 12	27	37
4, 5, 6 in 12	28	40

The spacing of trusses depends on the live load to be carried. A 24-in. spacing is recommended in most parts of the country. The trusses must be securely fastened to the bearing wall. Metal ties are recommended.

9.3.5 Post-and-Beam Framing

Post-and-beam or plank-and-beam framing is used in certain parts of the country where large glass areas are practical and where the architecture tends toward exposed ceiling beams and other exposed structural members. It is rarely used in conventional designs in other areas. The large timber posts and beams lend themselves to bridging over wide openings in both the exterior walls and in interior partitions. Such construction can, of course, be used in any part of the country when an effect of interior spaciousness and exterior light is required. It also lends itself to flat or slightly sloped roofs. When using this construction, the builder must observe certain precautions.

The exterior walls between large openings must be fully framed and braced to provide lateral resistance.

Heavy concentrated loads must be provided for by additional framing: for example, refrigerators, bathtubs, other heavy appliances, or furniture or files or books.

Because the underside of the roof planking is very often exposed as the ceiling, insulation must be applied over the planks and under the finish roofing. Two-inch-thick rigid foam is often used.

Because of the use of large framing members, the quality of the lumber must be good, the moisture content low, and the connections carefully made. Metal connections should be used.

Figures 9.37 and 9.38 show some of the details of the framing. It will be noted that this type of construction uses many fewer parts and is therefore labor-saving. However, heavy construction members cost more in proportion to lighter material and should be of better quality.

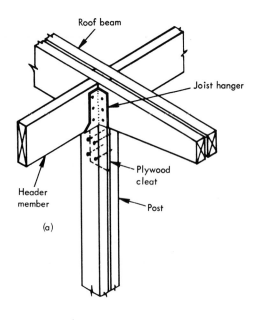

Roof beam

Joist hanger

Header member

Plywood cleat

Post

(a)

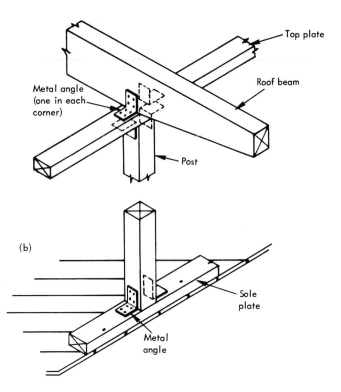

Top plate

Metal angle (one in each corner)

Roof beam

Post

(b)

Sole plate

Metal angle

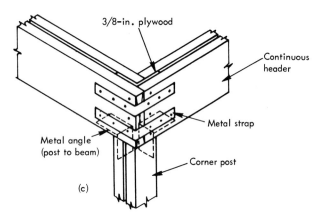

3/8-in. plywood

Continuous header

Metal strap

Metal angle (post to beam)

Corner post

(c)

Figure 9.37. Examples of post-and-beam construction: (a) to (c) connections at posts; (d) header and roof decking at exterior wall. Note the extensive use of metal connectors.

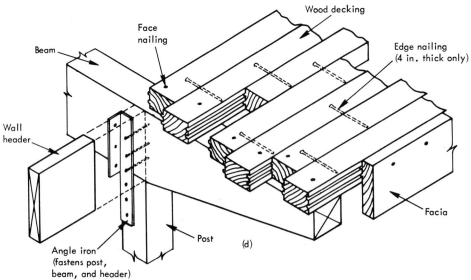

Wood decking

Face nailing

Edge nailing (4 in. thick only)

Beam

Wall header

Angle iron (fastens post, beam, and header)

Post

Facia

(d)

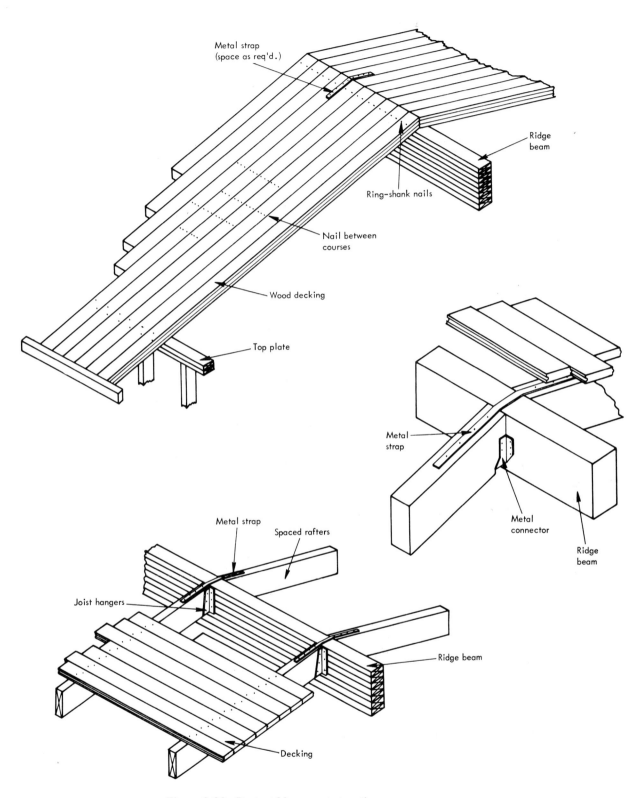

Figure 9.38. Post-and-beam construction:
Examples of decking and connections.

Metal strap
(space as req'd.)

Ridge
beam

Ring-shank nails

Nail between
courses

Wood decking

Top plate

Metal
strap

Metal
connector

Ridge
beam

Metal strap

Spaced rafters

Joist hangers

Ridge beam

Decking

153

9.4 COST ESTIMATING

The following sections will explain the basic knowledge that is required for the reliable estimating of the cost of a structural frame.

9.4.1 How to Measure a Board Foot

Building lumber comes in standard widths and thicknesses and in modular lengths, as explained in Section 9.2. The first step in ordering lumber after the materials list has been completed is to reduce the total quantity to board feet. All softwood framing lumber is ordered and priced by the thousand board feet (MBF). A board foot of lumber is 12 in. wide by 12 in. long by 1 in. thick. Board feet are measured by the nominal size and not the actual size (see Section 9.2). There are several ways of measuring board feet.

Thickness in inches by width in inches by length in feet:

2 in. × 10 in. × 6 ft long = 120 divided by 12 = 10 BF

Thickness in inches by width in inches by length in inches:

2 in. × 10 in. × 72 in. = 1440 divided by 144 = 10 BF:

A quick table for converting some standard sizes of framing lumber to board feet is:

2 x 4	Multiply linear feet by	0.66
2 x 6	Multiply linear feet by	1
2 x 8	Multiply linear feet by	1.33
2 x 10	Multiply linear feet by	1.66
2 x 12	Multiply linear feet by	2

There is also the essex board measure table, which is printed on the back of carpenter's framing squares. This converts lumber sizes to board feet by using a 1-in. thickness as a base. Table 9.7 gives the number of board feet for every size of framing lumber from 1 × 2 to 4 × 12 and for every length from 8 to 24 ft.

9.4.2 Quantity Takeoff and Materials List

For the purpose of this takeoff, a conventionally built one-story ranch house with a peaked roof will be taken as an example. The house is 27 ft 8 in. wide by 43 ft 4 in. long (Figure 9.39).

The first step is to use a form that will serve to remind the estimator of all the various members of the structure, so that the kind of material, the size, and the number of pieces to be used will be certain to be included. The list should show the number of board feet, lineal feet, or square feet, depending on how the material is sold. The materials list given in Figure 9.40 is confined to framing material only. Other materials will be listed in later chapters as they are shown.

TABLE 9.7. Board Feet in Standard Sizes of Framing Lumber

Nominal Size (in.)	\multicolumn{9}{c}{Actual Length (ft)}
	8	10	12	14	16	18	20	22	24
1 × 2		$1\frac{2}{3}$	2	$2\frac{1}{3}$	$2\frac{2}{3}$	3	$3\frac{1}{2}$	$3\frac{2}{3}$	4
1 × 3		$2\frac{1}{2}$	3	$3\frac{1}{2}$	4	$4\frac{1}{2}$	5	$5\frac{1}{2}$	6
1 × 4	$2\frac{3}{4}$	$3\frac{1}{3}$	4	$4\frac{2}{3}$	$5\frac{1}{3}$	6	$6\frac{2}{3}$	$7\frac{1}{3}$	8
1 × 5		$4\frac{1}{6}$	5	$5\frac{5}{6}$	$6\frac{2}{3}$	$7\frac{1}{2}$	$8\frac{1}{3}$	$9\frac{1}{6}$	10
1 × 6	4	5	6	7	8	9	10	11	12
1 × 7		$5\frac{5}{8}$	7	$8\frac{1}{6}$	$9\frac{1}{3}$	$10\frac{1}{2}$	$11\frac{2}{3}$	$12\frac{5}{6}$	14
1 × 8	$5\frac{1}{3}$	$6\frac{2}{3}$	8	$9\frac{1}{3}$	$10\frac{2}{3}$	12	$13\frac{1}{3}$	$14\frac{2}{3}$	16
1 × 10	$6\frac{2}{3}$	$8\frac{1}{3}$	10	$11\frac{2}{3}$	$13\frac{1}{3}$	15	$16\frac{2}{3}$	$18\frac{1}{3}$	20
1 × 12	8	10	12	14	16	18	20	22	24
$1\frac{1}{4}$ × 4		$4\frac{1}{6}$	5	$5\frac{5}{6}$	$6\frac{2}{3}$	$7\frac{1}{2}$	$8\frac{1}{3}$	$9\frac{1}{6}$	10
$1\frac{1}{4}$ × 6		$6\frac{1}{4}$	$7\frac{1}{2}$	$8\frac{3}{4}$	10	$11\frac{1}{4}$	$12\frac{1}{2}$	$13\frac{3}{4}$	15
$1\frac{1}{4}$ × 8		$8\frac{1}{3}$	10	$11\frac{2}{3}$	$13\frac{1}{3}$	15	$16\frac{2}{3}$	$18\frac{1}{3}$	20
$1\frac{1}{4}$ × 10		$10\frac{5}{12}$	$12\frac{1}{2}$	$14\frac{7}{12}$	$16\frac{2}{3}$	$18\frac{3}{4}$	$20\frac{5}{6}$	$22\frac{11}{12}$	25
$1\frac{1}{4}$ × 12		$12\frac{1}{2}$	15	$17\frac{1}{2}$	20	$22\frac{1}{2}$	25	$27\frac{1}{2}$	30
$1\frac{1}{2}$ × 4	4	5	6	7	8	9	10	11	12
$1\frac{1}{2}$ × 6	6	$7\frac{1}{2}$	9	$10\frac{1}{2}$	12	$13\frac{1}{2}$	15	$16\frac{1}{2}$	18
$1\frac{1}{2}$ × 8	8	10	12	14	16	18	20	22	24
$1\frac{1}{2}$ × 10	10	$12\frac{1}{2}$	15	$17\frac{1}{2}$	20	$22\frac{1}{2}$	25	$27\frac{1}{2}$	30
$1\frac{1}{2}$ × 12	12	15	18	21	24	27	30	33	36
2 × 4	$5\frac{1}{3}$	$6\frac{2}{3}$	8	$9\frac{1}{3}$	$10\frac{1}{3}$	12	$13\frac{1}{3}$	$14\frac{2}{3}$	16
2 × 6	8	10	12	14	16	18	20	22	24
2 × 8	$10\frac{2}{3}$	$13\frac{1}{3}$	16	$18\frac{2}{3}$	$21\frac{1}{3}$	24	$26\frac{2}{3}$	$29\frac{1}{3}$	32
2 × 10	$13\frac{1}{3}$	$16\frac{2}{3}$	20	$23\frac{1}{3}$	$26\frac{2}{3}$	30	$33\frac{1}{3}$	$36\frac{2}{3}$	40
2 × 12	16	20	24	28	32	36	40	44	48
3 × 6	12	15	18	21	24	27	30	33	36
3 × 8	16	20	24	28	32	36	40	44	48
3 × 10	20	25	30	35	40	45	50	55	60
3 × 12	24	30	36	42	48	54	60	66	72
4 × 4	$10\frac{2}{3}$	$13\frac{1}{3}$	16	$18\frac{2}{3}$	$21\frac{1}{3}$	24	$26\frac{2}{3}$	$29\frac{1}{3}$	32
4 × 6	16	20	24	28	32	36	40	44	48
4 × 8	$21\frac{1}{3}$	$26\frac{2}{3}$	32	$37\frac{1}{3}$	$42\frac{2}{3}$	48	$53\frac{1}{3}$	$58\frac{2}{3}$	64
4 × 10	$26\frac{2}{3}$	$33\frac{1}{3}$	40	$46\frac{2}{3}$	$53\frac{1}{3}$	60	$66\frac{2}{3}$	$73\frac{1}{3}$	80
4 × 12	32	40	48	56	64	72	80	88	96

Figure 9.39. One-story conventional house with a peaked roof.

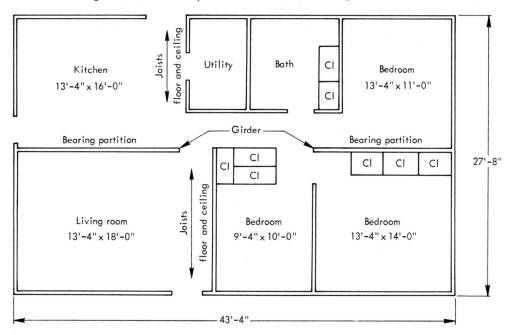

PROPERTY   DATE
_____

CONTRACTOR

Framing Member	Size	Length	Pieces	Total Length	Waste 10%	Board Feet
Sills	2X6	14	4	56		
	2X6	14	4	56	1-14	158
	2X6	16	2	32		
Girders	2X10	12	9	108	NONE	220
	2X10	8	3	24		
Floor Joists	2X8	14	75	1050		
Ceiling Joists					14-14s	3072
	2X8	14	75	1050		
Studs Exterior	2X4	8	150	1200		
Partition	2X4	8	150	1200	30-8s	1760
Plates	2X4	16	6	96		
	2X4	14	20	280	6-14s	552
	2X4	12	30	360		
Headers	2X8	4	14	56		
	2X8	2	16	32	NONE	120
Lintels	2X4	14	14	196	2-14s	152
Rafters	2X6	18	34	612	NONE	612

Figure 9.40. Materials list.

The list below indicates how the quantities shown in the materials list were obtained. There is a 10% allowance for waste in some items. If the builder takes some time to figure how best to use modular lengths, most of this waste can be eliminated.

Sills

The perimeter of the house is 2 x 43'-4" + 2 x 27'-8" = 142 ft. The sills are 2 x 6. The sides are 27'-8". Each side will require two 14-ft lengths, or 28 ft. The length is 43'-4". The front and rear will require one 16-ft length plus two 14-ft lengths (16 + 28), or 44 ft. There is almost no waste.

Girders

Girders will be three 2 x 10s. They have to span the length of 43'-4". There will be three 12-ft spans = 36 ft and one 8-ft span, or a total of 44 ft.

Floor and Ceiling Joists

The joists are 2 x 8s on 16-in. centers. A 14-ft joist will span a 13'-4" width of room and bear 4 in. on each side. The length of the house is

43'-4". Allow two joists per 16 in. (43.3 divided by 1.3 x 2), or 68, plus double joists at five parallel partition and double joists at gables, for a total of 68 + 5 + 2 = 75. Allow the same for ceiling joists to allow for future occupancy of attic space.

Studs

Studs are 2 x 4s on 16-in. centers. The perimeter of the house is 142 ft. Divide 142 by 1.3 = 110. Triple the studs at four corners and double them at 12 windows and two doors. Total is 110 + 12 + 28 = 150. The nearest modular length of studs is 8 ft.

Partitions

There are approximately 100 ft of partitions. With studs on 16-in. centers plus doubling at 13 doors and 22 corners equals 80 + 26 + 44 = 150.

Plates

The perimeter of the house is 142 ft and there is 100 ft of partition for a total of 242 ft. There are three 2 x 4 plates at all exterior and interior walls, for a total of 726 lineal feet. Plates should be staggered so that a mixture of different lengths is desirable. The 27'-8" sides can take 16- and 12-ft lengths which total 3 X 16 = 48 + 3 X 12 = 36, or 84 X 2 = 168 ft. The remaining 558 ft can be 20 14-ft and 24 12-ft lengths.

Headers

The front and rear total 88 ft. Each 44-ft side can take two 14-ft lengths = 28 + 1 - 16 = 44 X 2 = 88 ft.

Lintels

There are a total of 12 windows and 15 doors (interior and exterior). 27 openings X 3'-6" = 95 X 2 = 190 ft of 2 x 4, or 14 pieces 14-ft long.

Rafters

The slope of the roof should be sufficient to provide head room for living space in the attic floor. A rise of 10 ft at the peak will allow such room. The length of rafters will then be as follows:

The width 27.6 ÷ 2 = 13.8 X 13.8 = 190

The height   10   X 10   = 100

The length of the rafter is the square root of 190 + 100 = 290
    or 17 ft (from table or a computer)

Rafters are 2 x 6 and are on 16-in centers. 43.25 ÷ 1.3 = 34
    rafters

There are other framing members for dormers, collar beams, framing for openings, and so on. This calculation gives the essentials of the takeoff for a platform frame. Hardware must also be ordered. The nailing list gives the number and kind of nails to be used, and the chart of nail sizes gives the number of nails per pound. There also are bolts, hangers, structural fasteners, screws, and many other items to be included.

9.4.3   Pricing Materials and Labor Productivity

The bill of materials shown in Figure 9.40 is, as noted, for framing material only. Other material, such as rough and finish flooring, sheathing, roofing, and masonry, will be discussed in later chapters. However, the builder should gather all of the material lists into a single master list before requesting bids from a material dealer. At least two bids should be obtained and the builder should attempt to obtain as much of a breakdown of prices as possible. It is very possible that the sheathing material or roofing material or millwork may cost less from another dealer than from the one who is supplying the rough framing lumber. The builder should also make it understood that the price includes several shipments of material as the job progresses. If the price is right, the builder should also be somewhat flexible about the kind of material: for instance Douglas fir versus southern pine.

Labor productivity is difficult to measure unless the builder has had experience with the workers being employed on this job. There are tables of productivity prepared by the government which are used by the armed services. There are also several privately prepared surveys of productivity. All of these are based on average working conditions with regard to weather, skill, crew size, accessibility, and availability of standard-grade material. They are *average* figures. Following are some of these estimates.

		Source 1	*Source 2*
Girders	Three 2 X 8	40 labor-hours per MFBM	36 labor-hours
Joists	2 X 10	32 labor-hours per MFBM	24 labor-hours
Rafters	2 X 6	48 labor-hours per MFBM	24 labor-hours
Studs	2 X 4	Average of 56 labor-hours per MFBM	24 labor-hours
Plates	2 X 4	Average of 56 labor-hours per MFBM	40 labor-hours

As will be noted, the figures vary widely, but an average can reasonably be used. Wage rates vary all over the country, of course, so none are given here.

9.5.   PLYWOOD

Plywood is probably one of the most widely used wood construction materials. In its structural grades it is used for gusset plates, joist connectors, boxed beams, and for other uses where its strength under stress is required. In its standard grade it is used for subflooring, wall sheathing, and roof sheathing. Plywood comes in two structural grades—Structural 1 and 2—and in a Standard Grade. It also comes as Plyform for use in concrete formwork. These grades are marked on the sheets of plywood. It comes with exterior glue for use in areas where it is exposed to excessive moisture. There are interior grades for use in paneling, cabinets, decorative panels, and furniture.

Grade-use guide for construction and industrial plywood

	Use these symbols when you specify plywood (1) (2)	Description and Most Common Uses	Typical Grade-trademarks	Veneer Grade Face	Back	Inner Plys	1/4	5/16	3/8	1/2	5/8	3/4	7/8	1-1/8
Interior Type	STANDARD INT-DFPA (4)	Unsanded Interior sheathing grade for floors, walls and roofs. Limited exposure crates, bins, containers and pallets.	STANDARD 32/16 DFPA INTERIOR PS 1-66 000	C	D	D		■	■	■	■	■		
	STANDARD INT-DFPA (4) (with Exterior glue)	Same as Standard sheathing but has Exterior glue. For construction where unusual moisture conditions may be encountered. Often used for pallets, crates, bins, etc. that may be exposed to the weather.	STANDARD 32/16 DFPA INTERIOR EXTERIOR GLUE	C	D	D		■	■	■	■	■		
	STRUCTURAL I and STRUCTURAL II INT-DFPA	Unsanded structural grades where plywood design properties are of maximum importance. Structural diaphragms, box beams, gusset plates, stressed skin panels. Also for containers, pallets, bins. Made only with Exterior glue. Structural I limited to Group 1 species for face, back and inner plys. Structural II permits Group 1, 2, or 3 species.	STRUCTURAL I 32/16 DFPA INTERIOR EXTERIOR GLUE	C	D	D		■	■	■	■	■		
	UNDERLAYMENT INT-DFPA (4)	For underlayment or combination subfloor-underlayment under resilient floor coverings, carpeting. Used in homes, apartments, mobile homes, commercial buildings. Ply beneath face is C or better veneer. Sanded or touch-sanded as specified.	UNDERLAYMENT GROUP 2 DFPA INTERIOR PS 1-66 000	C Plugged	D	C & D	■		■	■	■	■		
	C-D PLUGGED INT-DFPA (4)	For utility built-ins, backing for wall and ceiling tile. Not a substitute for Underlayment. Ply beneath face permits D grade veneer. Also for cable reels, walkways, separator boards. Unsanded or touch-sanded as specified.	C-D PLUGGED GROUP 2 DFPA INTERIOR PS 1-66 000	C Plugged	D	D			■	■	■	■		
	2-4-1 INT-DFPA (5)	Combination subfloor-underlayment. Quality base for resilient floor coverings, carpeting, wood strip flooring. Use 2-4-1 with Exterior glue in areas subject to excessive moisture. Unsanded or touch-sanded as specified.	2-4-1 DFPA GROUP 2 INTERIOR PS 1-66 000	C Plugged	D	C & D								■
Exterior Type	C-C EXT-DFPA (4)	Unsanded grade with waterproof bond for subflooring and roof decking, siding on service and farm buildings. Backing, crating, pallets, pallet bins, cable reels.	C-C 32/16 DFPA EXTERIOR PS 1-66 000	C	C	C		■	■	■	■	■		
	C-C PLUGGED EXT-DFPA (4)	Use as a base for resilient floors and tile backing where unusual moisture conditions exist. For refrigerated or controlled atmosphere rooms. Also for pallets, fruit pallet bins, reusable cargo containers, tanks and boxcar and truck floors and linings. Sanded or touch-sanded as specified.	C-C PLUGGED GROUP 4 DFPA EXTERIOR PS 1-66 000	C Plugged	C	C	■		■	■	■	■		
	STRUCTURAL I C-C EXT-DFPA	For engineered applications in construction and industry where full Exterior type panels made with all Group 1 woods are required. Unsanded.	STRUCTURAL I C-C DFPA EXTERIOR PS 1-66 000	C	C	C		■	■	■	■	■		
	PLYFORM CLASS I & II B-B EXT-DFPA	Concrete form grades with high re-use factor. Sanded both sides. Edge-sealed and mill-oiled unless otherwise specified. Special restrictions on species. Also available in HDO.	B-B PLYFORM CLASS I EXTERIOR DFPA	B	B	C						■	■	

Notes:
(1) All Interior grades shown also available with Exterior glue.
(2) All grades except Plyform available tongue and grooved in panels 1/2" and thicker.
(3) Panels are standard 4x8-foot size. Other sizes available.
(4) Available in Group 1, 2, 3 or 4.
(5) Available in Group 1, 2, or 3 only.

Typical Back-stamp   ## Typical Edge-mark

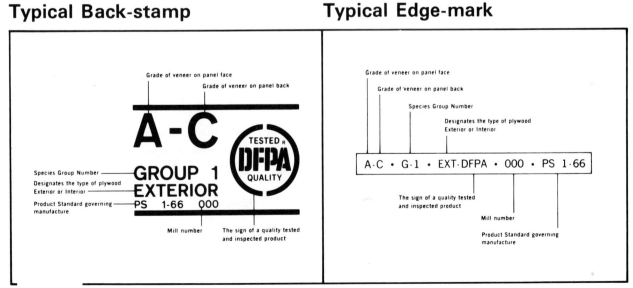

Figure 9.41. Grade-use guide for construction and industrial plywood. (Courtesy American Plywood Association)

Plywood Subflooring [1]

For application of 25 32" wood strip flooring or separate underlayment layer.

(Plywood continuous over 2 or more spans; grain of face plys across supports)

Panel Identification Index [2] [3]	Plywood Thickness (inch)	Maximum Span [4] (inches)	Nail Size & Type	Nail Spacing (inches)	
				Panel Edges	Intermediate
30/12	5/8	12 [5]	8d common	6	10
32/16	1/2, 5/8	16 [6]	8d common [7]	6	10
36/16	3/4	16 [6]	8d common	6	10
42/20	5/8, 3/4, 7/8	20 [6]	8d common	6	10
48/24	3/4, 7/8	24	8d common	6	10
2·4·1 and 1-1/8" Groups 1&2	1-1/8	48	10d common	6	6
1-1/4" Groups 3&4	1-1/4	48	10d common	6	6

[1] These values apply for Structural I and II, Standard sheathing and C-C Exterior grades only.

[2] Identification Index appears on all panels, except 1⅛" and 1¼" panels.

[3] In some non-residential buildings, special conditions may impose heavy concentrated loads and heavy traffic requiring sub-floor constructions in excess of these minimums.

[4] Edges shall be tongue and grooved, or supported with blocking, unless underlayment is installed, or finish floor is 25/32" wood strip. Spans limited to values shown because of possible effect of concentrated loads. At indicated maximum spans, panels will support uniform loads of at least 65 psf. For spans for 24" or less, panels will support loads of at least 100 psf.

[5] May be 16" if 25/32" wood strip flooring is installed at right angles to joists.

[6] May be 24" if 25/32" wood strip flooring is installed at right angles to joists.

[7] 6d common nail permitted if plywood is ½".

Figure 9.42. Plywood subflooring.
(Courtesy American Plywood Association)

Plywood Wall Sheathing [1]

(Plywood continuous over 2 or more spans)

Panel Identification Index	Panel Thickness (inch)	Maximum Stud Spacing (inches)		Nail Size [2]	Nail Spacing (Inches)	
		Exterior Covering Nailed to:			Panel Edges (when over framing)	Intermediate (each stud)
		Stud	Sheathing			
12/0, 16/0, 20/0,	5/16	16	16 [3]	6d	6	12
16/0, 20/0, 24/0	3/8	24	16 24 [3]	6d	6	12
24/0, 32/16	1/2	24	24	6d	6	12

[1] When plywood sheathing is used, building paper and diagonal wall bracing can be omitted.

[2] Common smooth, annular, spiral thread, or T-nails of the same diameter as common nails (0.113" dia. for 6d) may be used. Staples also permitted at reduced spacing.

[3] When sidings such as shingles are nailed only to the plywood sheathing, apply plywood with face grain across studs.

Figure 9.43. Plywood Wall Sheathing.
(Courtesy American Plywood Association)

In this book its use will be described for subfloors, wall sheathing, roof sheathing, structural reinforcement, and in a prefinished form for use as interior wall finishes.

The charts shown in Figures 9.41 to 9.45 were prepared by the American Plywood Association for the guidance of the builder. Figure 9.41 is a general-use chart. It shows the quality and the marking symbols. Figures 9.42 to 9.45 are guides to the thickness of the plywood to be used for various joist, stud, or rafter spans for subflooring, wall sheathing, and roof sheathing.

Plywood for exterior use comes in the following sizes, all ¾-in. thick only: 2 × 4, 4 × 4, 4 × 8, and 5 × 9.

Guide to Identification Indexes on plywood sheathing grades
(Based on Table 7, PS 1-66)

Identification Index Number	C-C Exterior and Standard Sheathing Thickness (inch)						Structural I Thickness (inch)						Structural II Thickness (inch)					
Roof Span / Floor Span	5/16	3/8	1/2	5/8	3/4	7/8	5/16	3/8	1/2	5/8	3/4	7/8	5/16	3/8	1/2	5/8	3/4	7/8
12/0	■																	
16/0	■	■											■					
20/0	■	■					■						■	■				
24/0		■	■					■						■	■			
30/12				■														
32/16			■	■					■						■	■		
36/16				■														
42/20				■	■	■				■						■	■	
48/24					■	■					■						■	■

Note:

The Identification Index numbers placed on the panel by the manufacturer in the grades shown in the table at left are based on panel thickness and the stiffness of species used on the face and back. Since stiffness varies with different species, the same Index number may appear on panels of different thickness. For example, 24/0 appears on both ⅜″ and ½″ Standard sheathing. In the latter, species of lesser stiffness have been used on the face and back. Conversely, panels of the same thickness may be marked with as many as three different Index numbers. For example, ⅜″ Standard may be marked with 16/0, 20/0 or 24/0 depending on the species used on the face and back. In general, the higher the Index number, the greater the stiffness.

Figure 9.44. Guide to identification indexes on plywood sheathing grades. (Courtesy American Plywood Association)

Plywood Roof Sheathing [1] [2] [3]

(Plywood continuous over 2 or more spans; grain of face plys across supports)

Panel Ident. Index	Plywood Thickness (inch)	Max. Span (inches) [4]	Unsupported Edge—Max. Length (inches) [5]	Allowable Roof Loads (psf) [6] [7]										
				(Spacing of Supports [Inches] Center to Center)										
				12	16	20	24	30	32	36	42	48	60	72
12/0	5/16	12	12	100 (130)										
16/0	5/16, 3/8	16	16	130 (170)	55 (75)									
20/0	5/16, 3/8	20	20		85 (110)	45 (55)								
24/0	3/8, 1/2	24	24		150 (160)	75 (100)	45 (60)							
30/12	5/8	30	26			145 (165)	85 (110)	40 (55)						
32/16	1/2, 5/8	32	28				90 (105)	45 (60)	40 (50)					
36/16	3/4	36	30				125 (145)	65 (85)	55 (70)	35 (50)				
42/20	5/8, 3/4, 7/8	42	32					80 (105)	65 (90)	45 (60)	35 (40)			
48/24	3/4, 7/8	48	36						105 (115)	75 (90)	55 (55)	40 (40)		
2-4-1	1-1/8	72	48							160 (160)	95 (95)	70 (70)	45 (45)	25 (30)
1-1/8" G 1&2	1-1/8	72	48							145 (145)	85 (85)	65 (65)	40 (40)	30 (30)
1-1/4" G 3&4	1-1/4	72	48							160 (165)	95 (95)	75 (75)	45 (45)	25 (35)

[1] Applies to Standard, Structural I and II and C-C grades only.

[2] For applications where the roofing is to guaranteed by a performance bond, recommendations may differ somewhat from these values. Contact American Plywood Association for bonded roof recommendations.

[3] Use 6d common smooth, ring-shank or spiral thread nails for ½" thick or less, and 8d common smooth, ring-shank or spiral thread for plywood 1" thick or less (if ring-shank or spiral thread nails same diameter as common). Use 8d ring-shank or spiral thread or 10d common smooth shank nails for 2-4-1, 1⅛" and 1¼" panels. Space nails 6" at panel edges and 12" at intermediate supports except that where spans are 48" or more, nails shall be 6" o.c. at all supports.

[4] The spans shall not be exceeded for any load conditions.

[5] Provide adequate blocking, tongue and grooved edges or other suitable edge support such as PlyClips when spans exceed indicated value. Use two PlyClips for 48" or greater spans and one for lesser spans.

[6] Uniform load deflection limitation: 1/180th of the span under live load plus load, 1/240th under live load only. Allowable live load shown in boldface type and allowable total load shown within parenthesis.

[7] Allowable loads were established by laboratory test and calculations assuming evenly distributed loads. Figures shown are not applicable for concentrated loads.

Figure 9.45. Plywood roof sheathing. (Courtesy American Plywood Association)

CHAPTER TEN

exterior wall cladding

10.1 THE CHOICE OF MATERIALS

A viewer's first impression of a house is gained from its exterior appearance. The exterior material and its mode of installation can be used to express the architecture and to an extent a way of life. It is therefore most important for the owner and/or builder to choose a material and installation that will make the house appealing in the general area and the neighborhood in which it is located.

In many areas, especially in the more-northern climate, a house covered with wood siding is most widely used. There are many kinds of such siding. Wood-frame houses are also covered with masonry or stone veneer, prefinished metal siding, plastic siding, wood shingles, and stucco, which is used in the more temperate parts of the country. Certain grades of plywood, fiberboard, and hardboard are also sometimes used.

10.2 SHEATHING

The exterior framed walls must be braced or covered with a layer of material underneath the exterior wall finish.

10.2.1 Wood Sheathing

The sheathing may consist of boards which may be square-edged or tongue-and-groove. Such boards usually have a nominal thickness of 1 in. and a width of 6, 8, or 10 in. Sheathing can be placed horizontally, or diagonally,

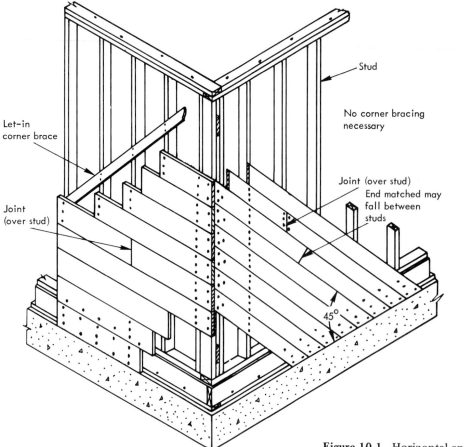

Figure 10.1. Horizontal and diagonal wood sheathing.

but if placed horizontally it is recommended that corner bracing be used. As a matter of fact, corner bracing is required by most codes when nonrigid sheathing is used. Diagonal sheathing serves as bracing. The material of board sheathing is usually of economy-grade spruce or hemlock, but Douglas fir and southern pine can also be used. To prevent warping and open joints the moisture content should be kept to 15%. Figure 10.1 shows the use of wood sheathing. The nailing schedule in Section 9.2 gives the number and size of nails that should be used.

10.2.2 Plywood Sheathing

Plywood sheathing is very often used for wall sheathing, especially in more severe climates. Because of its structural strength and because it comes in large sheets that can be nailed to a number of studs, it requires no corner bracing. The standard grades of plywood are used in most cases, but if there are severe climatic conditions the plywood can be obtained with a moisture-resistant exterior glueline. The tables in Section 9.5 give recommended thicknesses and qualities of plywood for various purposes. Figure 10.2 shows the installation of plywood as well as of structural insulating board, as described in the following section.

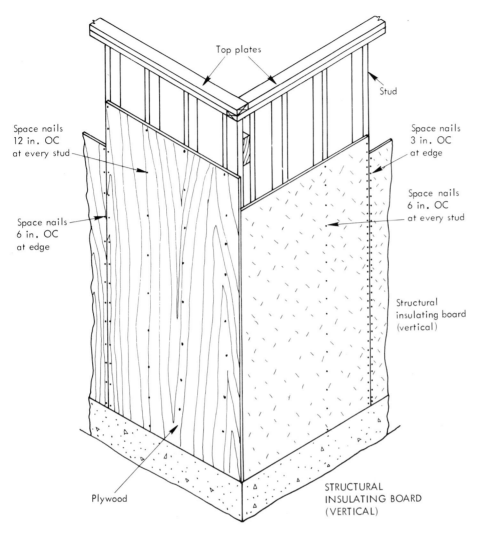

Top plates

Stud

Space nails
12 in. OC
at every stud

Space nails
3 in. OC
at edge

Space nails
6 in. OC
at every stud

Space nails
6 in. OC
at edge

Structural
insulating board
(vertical)

Plywood

STRUCTURAL
INSULATING BOARD
(VERTICAL)

Figure 10.2. Plywood and structural insulating board sheathing.

10.2.3   Structural Insulating Board and Gypsum Sheathing

Structural insulating board can be used as sheathing. It is made of compressed wood fiber and is impregnated with asphalt to make it waterproof. If the siding or other exterior finish is to be nailed directly to it, the board must be of "nail-base" quality. Corner bracing is required for all of this type of sheathing when it is laid horizontally and always on the ½-in. regular-density board. Insulating board comes in sizes 2 x 8, 4 x 8, and 4 x 9. It is recommended that galvanized or other corrosion-resistant nails be used. The board must be nailed to every stud.

   Gypsum sheathing is made of gypsum faced on both sides with water-resistant paper. It comes in 2 x 8 sheets, is applied horizontally, and requires corner bracing. As shown in Figure 10.3, gypsum board sheathing requires nailing strips to hold the exterior finish.

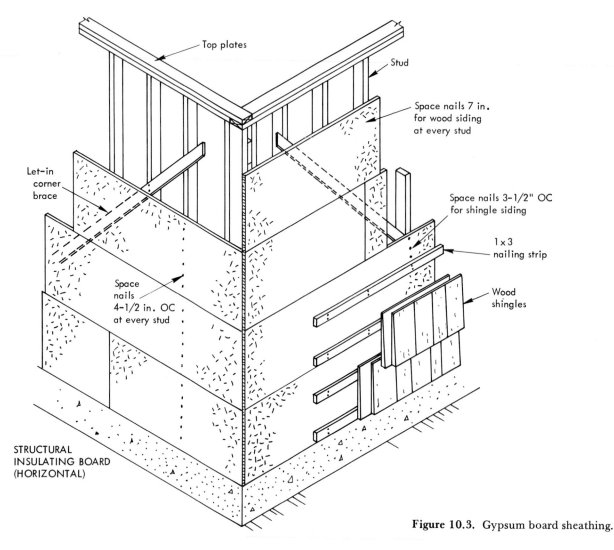

Top plates

Stud

Space nails 7 in.
for wood siding
at every stud

Let-in
corner
brace

Space nails 3-1/2" OC
for shingle siding

1 x 3
nailing strip

Space
nails
4-1/2 in. OC
at every stud

Wood
shingles

STRUCTURAL
INSULATING BOARD
(HORIZONTAL)

Figure 10.3. Gypsum board sheathing.

10.2.4 Sheathing Paper

Sheathing paper is water-resistant and is used only where this quality is required. There are sheathing papers with a central core made of fiber laid in asphalt or resin to make them resistant to tearing. Many builders use 15-lb asphalt felt. It is not considered necessary in most climates to use sheathing paper over plywood, fiberboard, or other sheathing material that is dense and water-resistant. It should be used over wood sheathing or when stucco or masonry veneer are used for the exterior finish.

10.3 EXTERIOR WALLS

The exterior wall is what is seen first. It should therefore not only be in keeping with the architectural character of the house but should also be as maintenance-free as possible. It must be structurally sturdy, resistant to weather, resistant to hard knocks, and have some insulating value.

10.3.1 Wood

Wood siding comes in many shapes and sizes. It should be easy to work, be resistant to warping, and have good painting qualities. It should have such moisture content that it will not dry further after it is installed and thus shrink or warp. It should be free of knots or pitch pockets. Normal moisture content for most of the country should not exceed 10 to 12%; in drier portions of the country, such as the Southwest, it should not exceed 8 to 9%.

The woods that are best for this purpose are cedar, eastern and western white pine, cypress, and redwood. Western hemlock, ponderosa pine, spruce, or yellow poplar are also used. Edge grain or vertical grain should be used to minimize shrinkage or swelling. Many manufacturers dip their siding in a water-repellent preservative. This is highly recommended as an item of preventive maintenance. It prevents decay and enhances paint adherence.

Horizontal Siding

The shapes that are most frequently used for horizontal siding are bevel siding, drop siding, anzac bevel siding, and paneling (Figure 10.4).

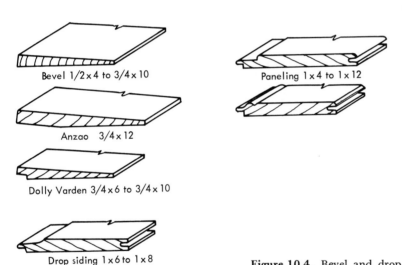

Bevel 1/2 x 4 to 3/4 x 10

Anzao 3/4 x 12

Dolly Varden 3/4 x 6 to 3/4 x 10

Drop siding 1 x 6 to 1 x 8

Paneling 1 x 4 to 1 x 12

Figure 10.4. Bevel and drop siding horizontal only; panel siding both horizontal and vertical.

Bevel siding should be lapped 2 in. to provide for resistance to weather and any shrinkage. It can be laid with smooth or rough-sawn side out. It comes in nominal sizes ½ x 4, ½ x 6, and ½ x 8, and also ¾ x 8 and ¾ x 10. The amount of the exposed area of bevel siding must be carefully figured so that the exposed area of each board is equal. A good way to determine this width of exposure is to determine the size of the window frame from bottom of sill to top of drip cap and then to divide this evenly as follows. Bottom of sill to top of drip cap is 5 ft 2 in. Using 8-in. siding with 6-in. exposure, divide 62 in. by 6 = 10²⁄₆ pieces. The ²⁄₆ of 6 in. is 2 in. This 2 in. can be divided by 10 pieces so that each piece will be

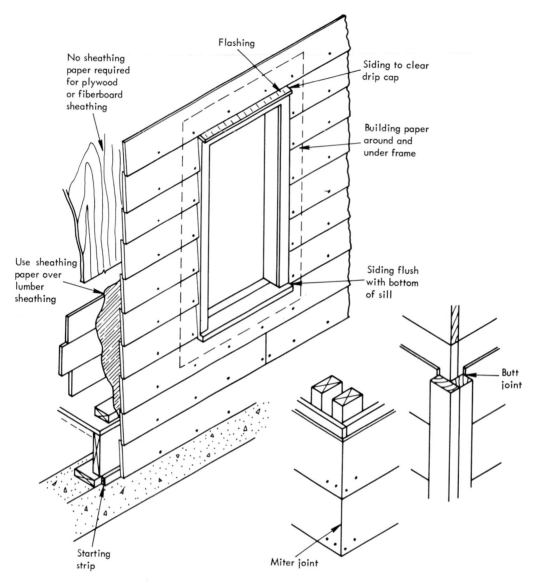

No sheathing paper required for plywood or fiberboard sheathing

Flashing

Siding to clear drip cap

Building paper around and under frame

Use sheathing paper over lumber sheathing

Siding flush with bottom of sill

Butt joint

Starting strip

Miter joint

Figure 10.5.
Bevel siding and corner treatment.

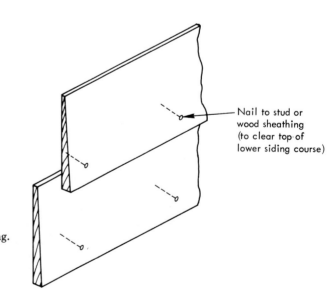

Nail to stud or wood sheathing (to clear top of lower siding course)

Figure 10.6.
Nailing of bevel siding.

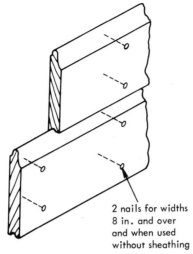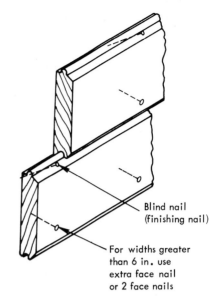

2 nails for widths
8 in. and over
and when used
without sheathing

Blind nail
(finishing nail)

For widths greater
than 6 in. use
extra face nail
or 2 face nails

Figure 10.7.
Drop or rabbeted siding paneling.

exposed 6.2 in. If 6-in. siding is used, it will be $62 \div 4 = 15\frac{1}{2}$ pieces exposed 4 in. The 2 in. left over can be divided among the 15 pieces. Figure 10.5 shows how bevel siding is measured and Figure 10.6 shows how it is applied. Nails should be corrosion-resistant and driven only flush with the surface, to avoid hammer marks. It is also recommended that annular threaded or helically threaded shank nails be used. The joists should be staggered and butt ends should mostly be used at openings.

Anzac and Dolly Varden bevel siding are also used as horizontal siding. Anzac comes only in $\frac{3}{4}$ x 12 dimensions. Dolly Varden is simply bevel siding with shiplap edges so that every exposed surface is equal. It lies flat against the studs and can be used for accessory buildings without sheathing but with corner bracing.

Drop siding comes in several patterns, all of which are essentially tongue-and-groove (T&G). It comes with shiplap edges in 1 x 6 and 1 x 8. Some patterns are known as "paneling." These come with a decorative bead or a rounded edge so that they form a horizontal (or a vertical) pattern. Figure 10.7 shows how drop siding is applied. Because it is T&G, it has a constant width of exposed surface. Although the T&G feature is more resistant to weather, drop siding is not normally used in better-quality houses. The nailing is the same as for bevel siding.

TABLE 10.1. Bevel Siding

Nominal Size (in.)	Butt Thickness (in.)	Actual Width (in.)	Quantity Required (ft²)
$\frac{1}{2}$ × 4	$\frac{15}{32}$	$3\frac{1}{2}$	1600
$\frac{1}{2}$ × 5	$\frac{15}{32}$	$4\frac{1}{2}$	1428
$\frac{1}{2}$ × 6	$\frac{15}{32}$	$5\frac{1}{2}$	1333
$\frac{5}{8}$ × 10	$\frac{9}{16}$	$9\frac{1}{2}$	1177
$\frac{3}{4}$ × 8	$\frac{3}{4}$	$7\frac{1}{2}$	1231
$\frac{3}{4}$ × 10	$\frac{3}{4}$	$9\frac{1}{2}$	1177
$\frac{3}{4}$ × 12	$\frac{3}{4}$	$11\frac{1}{2}$	1143

Tables 10.1 and 10.2 show how many square feet of bevel and panel siding are required to cover 1000 ft² of surface.

The quantities given in Table 10.1 are calculated on 1-in. lap. An allowance of 10% additional should be made for cutting and fitting in both Tables 10.1 and 10.2.

TABLE 10.2. Panel Siding

Nominal Size (in.)	Thickness (in.)	Quantity Required (ft²)
1 × 4	¾	1231
1 × 5	¾	1177
1 × 6	¾	1143
1 × 8	¾	1104
1 × 10	¾	1084

Shingles and Shakes

The use of shingles or shakes (which are a rustic or rough form of shingles) are very suitable for use in certain types of architecture. They look best on Colonial peaked-roof houses or the Cape Cod verson of Colonial. The shingle material that is most often used is western red cedar. This comes in several grades, lengths, and butt thicknesses. The thicker butt is used for shadow lines and an interesting architectural effect. It is also sturdier. Shingles come in bundles and the table that follows shows how much face area is covered per bundle in good-quality construction.

Type	Coverage
16-in. best grade	35 ft² at 7-in. exposure
18-in. best grade	39 ft² at 8½-in. exposure
18-in. red grade	39 ft² at 8½-in. exposure
24-in. best grade	
No. 1 Royals	35 ft² at 10-in. exposure

Shakes of hand-split and red cedar come in 18-, 24-, and 32-in. lengths. They come in butt thicknesses of ⅜ to ⅝, ½ to ¾, and ¾ to 1¼. They can be applied with the recommended exposures given in Table 10.3. These exposures are maximum requirements that cannot be exceeded.

Shingles and shakes can be nailed directly to wood or plywood sheathing. Nailing strips are required for all other sheathing. Figures 10.8 and 10.9 show their application in single or double coursing. Double coursing is high-quality construction and is not often used at current high construction costs. Nailing should be done with corrosion-resistant nails. The zinc-coated shingle nail is normally used in single coursing where nails are concealed. In double coursing exposed nails should be a 5d flat-head zinc-coated nail.

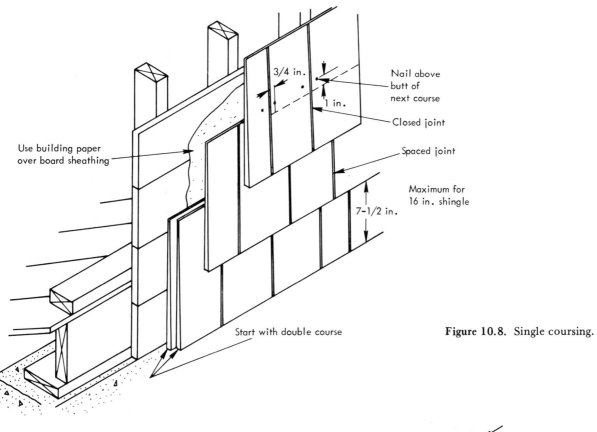

Use building paper over board sheathing

3/4 in.

Nail above butt of next course

1 in.

Closed joint

Spaced joint

Maximum for 16 in. shingle

7-1/2 in.

Start with double course

Figure 10.8. Single coursing.

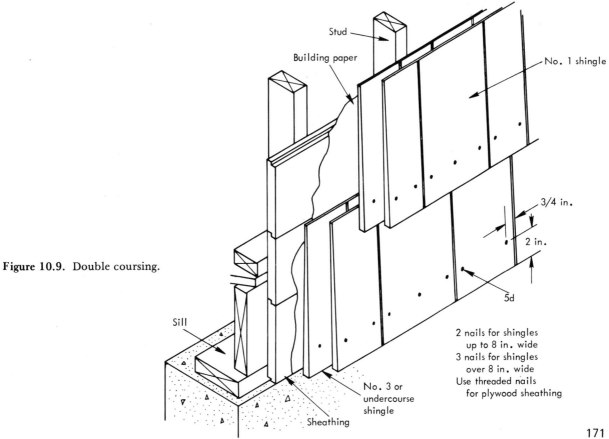

Stud

Building paper

No. 1 shingle

3/4 in.

2 in.

5d

Figure 10.9. Double coursing.

Sill

No. 3 or undercourse shingle

Sheathing

2 nails for shingles up to 8 in. wide
3 nails for shingles over 8 in. wide
Use threaded nails for plywood sheathing

171

TABLE 10.3. Recommended Exposures for Shingles and Shakes

	Length (in.)	Maximum Exposures (in.)	
		Single Course	Double Course
Shingles	16	7½	12
	18	8½	14
	24	11½	16
Shakes	18	8½	14
	24	11½	20
	32	15	

Vertical Siding

Vertical siding is used where certain architectural effects are required, such as in many contemporary styles and in cases where a rustic look is wanted. It comes in various patterns such as T&G with and without a V-notch, board and batten, board and board, batten and board, and channel rustic paneling. Most wood vertical siding is made of red cedar, although pine and fir are also used. The V-notched, T&G, and paneling are usually smooth-surfaced. The others are rough-sawn or sawn-textured. Some of the patterns are shown in Figure 10.10.

Vertical siding is applied in the same manner as horizontal siding. It can be nailed directly to wood or plywood sheathing and to nailing strips when other sheathing is used. It is recommended that nailing strips be no more than 24 in. apart, and 16 in. is better. Board-and-batten-type sidings should be nailed to nailing strips even when plywood or wood sheathing is used. Because vertical siding is vulnerable to wind-driven rain and weather in general, it is recommended that sheathing paper be used under it.

Figure 10.10. Types of vertical siding: (a) board and batten; (b) board and board; (c) batten and board; (d) channel rustic.

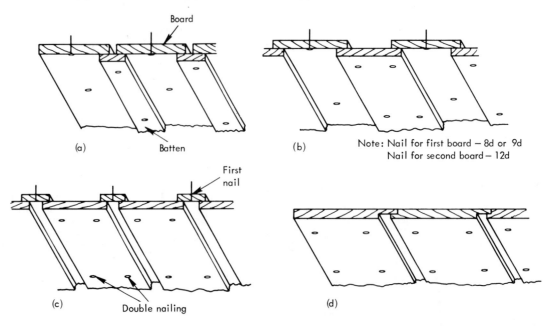

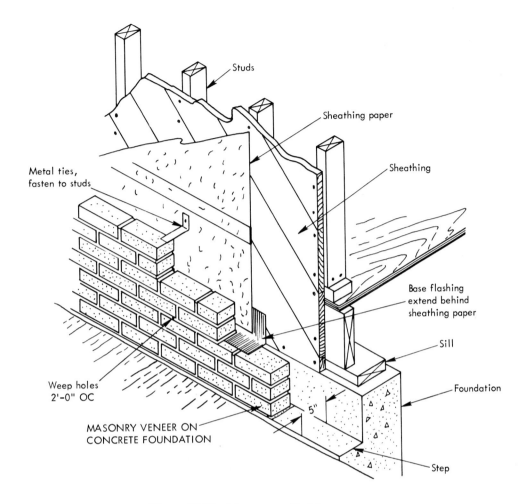

Figure 10.11. Masonry and brick veneers.

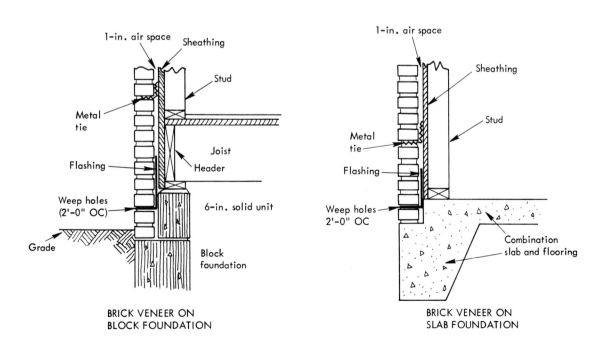

Labels for top figure:

Studs

Sheathing paper

Sheathing

Metal ties, fasten to studs

Base flashing extend behind sheathing paper

Sill

Foundation

Weep holes 2'-0" OC

5"

Step

MASONRY VENEER ON CONCRETE FOUNDATION

Labels for bottom-left figure:

1-in. air space

Sheathing

Stud

Metal tie

Joist

Header

Flashing

Weep holes (2'-0" OC)

6-in. solid unit

Grade

Block foundation

BRICK VENEER ON BLOCK FOUNDATION

Labels for bottom-right figure:

1-in. air space

Sheathing

Stud

Metal tie

Flashing

Weep holes 2'-0" OC

Combination slab and flooring

BRICK VENEER ON SLAB FOUNDATION

10.3.2 Masonry

Brick veneer over wood-frame exterior walls is used in many areas of the country. The usual face brick is made of selected clays and shales and is hard-burned. It is quite impermeable to weather and is maintenance-free. Any moisture penetration in a brick wall is usually through poorly laid joints. The brick should be laid with full bed joints and full cross joints, and careful attention must be paid to the composition of the mortar. Brick mortars are manufactured by companies which do not normally disclose their proprietary formulas. The basic purpose of the mortar is not only to form a bond between the bricks but also to form a workable mixture that will harden into a waterproof, impermeable material. Most mortars are made of a mixture of portland cement and various quantities of lime, and this mixture is, in turn, mixed with a set amount of sand and water. Masons often make their own mortar mix in proportions that they have found to be most suitable.

If it is planned to use a brick veneer wall, the builder must make provisions for it by placing a step in the foundation wall to support its weight, as shown in Figure 10.11, which shows the brick supported by a concrete wall, a block wall, and a foundation for a flat slab.

The brick must be attached to the frame wall by means of metal anchors set at distances as called for by local codes. Sheathing paper must be used over the sheathing. Two cautions must be observed: (1) to keep the space between the brick and the frame wall free of mortar droppings, and (2) to be careful to prevent the mortar from freezing in cold weather. This can be done by heated mortar, admixtures, and proper protection until the mortar sets. Tooling of the joints is also recommended.

A brick masonry veneer wall can be made quite attractive by the use of good-quality hard-surfaced face brick which comes in many color variations. The pattern bond in which the brick is laid can also add to the interest. Figure 10.12 shows three examples of pattern bonding. The builder should be familiar with certain common terms that are used in brick masonry. Figure 10.13 illustrates some of these terms.

The quantities of brick and mortar that are required for 4-in. brick veneer walls and in 8-in. bearing walls are shown in Table 10.4.

Single-family residences are rarely built of solid masonry 8-in. bearing walls. Such walls are more often used in low multiple dwellings of the "garden" type and in low office buildings. In such cases the floor framing may be of light steel joists (expanded joists) and mesh-reinforced concrete slabs.

Figure 10.12. Samples of pattern bonding:
(a) common; (b) running; (c) flemish.

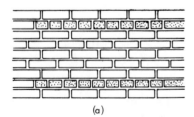

(a)

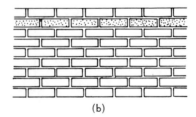

(b)

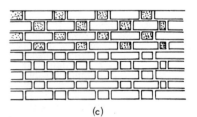

(c)

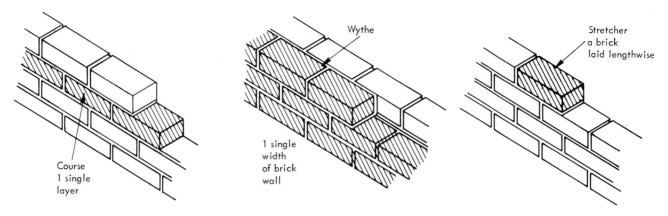

Figure 10.13. Common terms used in brick masonry.

TABLE 10.4. Brick and Mortar Quantities for Selected Wall Areas

| Wall Area (ft²) | Brick and Mortar Required[a] | | | |
| | 4 in. | | 8 in. | |
	Number of Bricks	Cubic feet of Mortar	Number of Bricks	Cubic feet of Mortar
1	6.17	0.08	12.33	0.2
10	61.7	0.8	123.3	2
100	617	8	1,233	20
200	1,234	16	2,466	40
300	1,851	24	3,699	60
400	2,468	32	4,932	80
500	3,085	40	6,165	100
600	3,712	48	7,398	120
700	4,319	56	8,631	140
800	4,936	64	9,864	160
900	5,553	72	10,970	180
1,000	6,170	80	12,330	200

[a]Quantities are for ½-in. mortar joints. For ⅜ in., use 80%; for ⅝ in., use 120%.

10.3.3   Metal

One of the most popular of the so-called maintenance-free materials that is used for exterior wall cladding is aluminum siding. Except by close inspection, this material is almost indistinguishable from wood siding. It should be of heavy-enough gauge to resist denting by a casual blow. It should have a baked-on acrylic finish over a prepared surface. The builder is warned to purchase a well-known brand product by a company that actually manufactures the siding. (Some companies just sell the sheet aluminum.)

The siding is applied over sheathing and building paper. The insulation value of aluminum siding is less than that of wood, but exterior wall cladding is not normally chosen for its insulating quality. Some aluminum comes with an insulating polystyrene backing, but the ⅜-in. thickness should not be depended on for much insulation value. From the standpoint of ease of maintenance, good-quality aluminum siding may last 20 or more years without painting. Aluminum siding also comes in textured patterns to imitate shingles and shakes, and as vertical siding.

10.3.4   Stucco

Stucco is a cement, lime, and sand mixture which is mixed with water to form a plastic mortar that is used as an exterior building coating. It is very popular in the milder climates of the country, where it will not be subject to severe weather. It can be colored and textured as it is applied for pleasing architectural effects. Stucco can be applied over wood frame or masonry walls. Figure 10.14 shows one of the ways of applying stucco to a wood frame wall. Water-repellent sheathing paper is used over wood sheathing.

A standard formula for stucco over masonry is 1 bag of portland cement, 1 bag of hydrated lime, and 6 ft^3 sand for a first coat, and 1 bag of portland cement, 10 lb of lime putty, and 3 ft^3 of moist sand for a second coat, or the latter formula as a single coat over wire lath. It is not advisable to use stucco as an exterior covering in severe climates. Water will enter any joint or crack, and when it freezes it will tend to widen the crack and split the stucco. In ordinary use, stucco is applied to wire lath held by furring strips over sheathing and sheathing paper. It is applied directly to dampened masonry.

10.3.5   Other Exterior Cladding Material

In addition to the foregoing, there are several other materials that are used for outer-wall cladding. The use of many of them depends on the area, the climate, and the architectural effect that is wanted. Plywood that is suitable for exterior use can be installed in vertical panels. It comes in various finishes

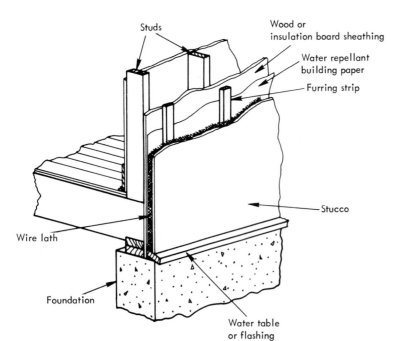

Figure 10.14. Stucco over wood framing.

and textures. Asbestos-cement-textured siding and shingles are often used in severe climates and for ease of maintenance. Hardboard is an impregnated-wood-fiber material that is bonded under heavy pressure and is also used in place of wood. The hardboard can be obtained in large panels with a choice of color and texture, or as siding in $7/16$ in. thickness by 16 ft long and in three widths: 6, 9, and 12 in. It is applied over sheathing and sheathing paper and usually on nailing strips. The advantage of hardboard or asbestos cement is a grainless structure, no corrosion, resistance to weather, and no painting required. There is also vinyl siding, which is tough, weather resistant, and maintenance free.

10.4 COST ESTIMATING

Before attempting to estimate the cost of cladding, the builder or owner must first determine the quality of the construction. In cases in which an architect is involved, the choice of materials and architecture are made by the owner and architect; the builder only estimates the cost of the material and method of construction as specified. The builder who is also the owner, as in speculative building, can choose what will appeal most to a prospective purchaser or what he decides on the basis of personal choice. The considerations involved in choosing materials are the first cost of the material; the cost of labor to install it; its resistance to weather, decay, and corrosion; and the ease and cost of maintaining it.

10.4.1 Material

In estimating the cost of material for exterior wall cladding, the first step is to establish how many square feet of surface are to be covered. Tables 10.1 and 10.2 show how much material is required to cover 1000 ft² of *exposed* surface. The exposed-surface measurement is especially important for all materials that lap, such as shingles, shakes, bevel siding, and some vertical siding. Unit prices for all these materials can be obtained from the local material dealer based on a list of materials.

A survey of comparative prices for sheathing and finish cladding should be made. It is suggested that quoted prices be set up in a table, as follows:

	Cost per 100 ft²
Sheathing	
$3/8$-in. plywood	
$1/2$-in. plywood	
Structural insulating board	
1 X 8 T&G boards	
Siding—Horizontal	
$1/2$ X 6 beveled cedar	
$1/2$ X 8 beveled cedar	
Shingles, 16-in. cedar	
Siding—Vertical	
$5/8$ X 8 grooved rough-sawn cedar	
$7/16$-in. grooved panel (Masonite)	

10.4.2 Labor

It should be obvious that the cost of labor bears a direct relationship to the material that is being installed. In the following figures, the average national wage rate of a carpenter is taken at $14.00 per hour, which includes benefits as set by union agreements. These are taken from a national guide on estimating. The builder must adjust the unit prices as shown to the local wage rates and to the productivity of the crew. The builder must also remember that wage rates change at each contract period. The prices that are given below represent the cost of installing 100 ft² of the various materials.

Sheathing	Labor Cost	Siding	Labor Cost
⅜-in. plywood	$19	½ × 6 beveled cedar	$30
½-in. plywood	20	½ × 8 beveled cedar	27
Structural insulating board	23	Shingles, 16-in. cedar	34
1 × 8 T&G boards	17	⅝ × 8 rough-sawn siding	34
		¾ × 8 shiplap	27

Although these figures should be used only as a guide, they do show how an estimate is put together: the determination of the quality, the breakdown into units, and the cost of labor and material per unit.

CHAPTER ELEVEN

roofing, flashing, gutters, and downspouts

11.1 FACTORS IN THE CHOICE OF MATERIAL

This chapter is about the weatherproofing of a house. Roofing, flashing and sheet metal in this capacity are used not only to keep rain, snow, sleet, and wind from entering the interior, but also to lead it away in as rapid a manner as possible to prevent it from building up and doing damage. Roofing and flashing act as barriers to weather; leaders and gutters are used to lead moisture away. The owner, architect, and builder have a wide choice of materials for all these purposes. There are several factors to consider in choosing the proper material: the ability of the material (1) to give long-lasting, maintenance-free service; (2) to fit the architecture of the house; (3) to be aesthetically pleasing; and (4) to be economical.

11.2 ROOF COVERING

Roof covering can consist of wood shingles or shakes or asphalt shingles for the usual single-family residence with a peaked roof. There are also mineral-fiber shingles, asbestos shingles, glass-fiber shingles, slate, clay tile, and cement tile. The latter two coverings are usually used only in warmer climates.

Built-up roofing is used for flat or slightly sloped roofs and consists of several layers of impregnated roofer's felt that is laid in moppings of hot tar or asphalt. It is not recommended that this type of roof be installed by other than a professional who has the proper equipment and the expertise.

Metal roofing is sometimes used on roofs over dormers, entryways, or porches. The metal can be copper or aluminum or other rust-resistant material. Here again, as for built-up roofing, this material must be carefully applied.

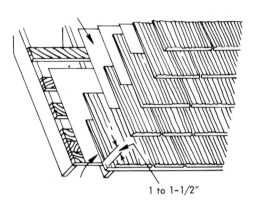

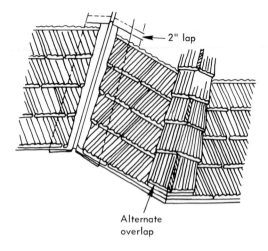

Figure 11.1. Application of roof shakes.

11.2.1 Wood Shingles and Shakes

The wood shingles that are used for roof covering should be of No. 1 grade and usually consist of the heartwood of red cedar or redwood. This is the part of the tree that is most resistant to decay and shrinkage. Wood shingles are also available made of cypress or white cedar. They are all edge grain and taper split. The most commonly available shingles are made of red cedar and come in 16-, 18-, and 24-in. lengths and in random widths. The thickness of the shingle butts varies from $\frac{3}{8}$ in. for the 16-in. shingle to $\frac{1}{2}$ or $\frac{5}{8}$ in. in the 18- and 24-in. lengths.

Shakes are rougher and more rustic looking than shingles and should be used with some regard for the architecture and location of the house. (Figure 11.1). They come as handsplit and resawn or as tapersplit or barn shakes. For roofing they are available in 18- and 24-in. lengths and in butt thick-

TABLE 11.1. Recommended Exposure and Coverage Areas for Shingles and Shakes

Shingle Length (in.)	Shake Length (in.)	Butt Thickness (in.)	Maximum Exposure (in.) 3–4 in 12	Maximum Exposure (in.) 5 or Over	Coverage (ft²) of 1 bundle at 5 in.
16		$\frac{3}{8}$	$3\frac{3}{4}$	5	25
18		$\frac{1}{2}$–$\frac{5}{8}$	$4\frac{1}{4}$	$5\frac{1}{2}$	25
24		$\frac{1}{2}$–$\frac{5}{8}$	$5\frac{3}{4}$	$7\frac{1}{2}$	25
	18	$\frac{3}{8}$		$5\frac{1}{2}$	25
	24	$\frac{1}{2}$–$1\frac{1}{4}$		10	25
	32	$\frac{3}{4}$–$1\frac{1}{4}$		13	25

nesses of ½ to ¾ in. or ¾ to 1¼ in. for the handsplit and resawn, in ½ in. for tapersplits, and in ⅜ in. for the 18-in. barn shakes.

Shingles and shakes are not recommended for use when the roof slope for the main house is less than 4 ft in 12 and for use over porches or other extensions when the slope is less than 3 ft in 12. Table 11.1 shows recommended exposures and the area that is covered.

Shingles and shakes are applied over roof boards or over furring strips. No felt underlayment is necessary except that in severe climates it is recommended that a 36-in. width of heavy roofing felt (30 to 45 lb) be rolled under the first few courses to serve as a flashing and a protection against an ice or snow dam at the gutter line. Such a dam will cause melting snow or ice to back up under the shingles (Figure 11.2).

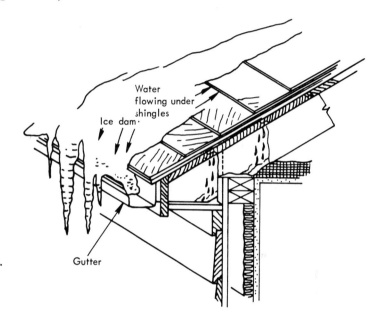

Figure 11.2.
Formation of an ice dam in a gutter.

The following procedure is recommended in the application of wood shingles:

Double the first course. The first course should extend 1½ in. beyond the eave line so as to carry water over the width of the fascia board. Shingles should also extend at least ¾ in. beyond the gable end.

The next and following course are laid on a line to give the proper exposure to the lower courses. Rust-resistant nails are used. There should be two nails to each shingle, spaced ¾ in. from the edge and 1½ in. beyond the butt line of the next course: 3d nails for 16- and 18-in. shingles and 4d for 24 in.

Shingles should be separated by about a ³⁄₁₆-in. space to allow for swelling, and the joints between shingles of any course should be offset at least 1½ in. (more is better) from the joints of the next course below (Figure 11.3).

Use the widest shingles as the first course away from valleys, and accurately precut them.

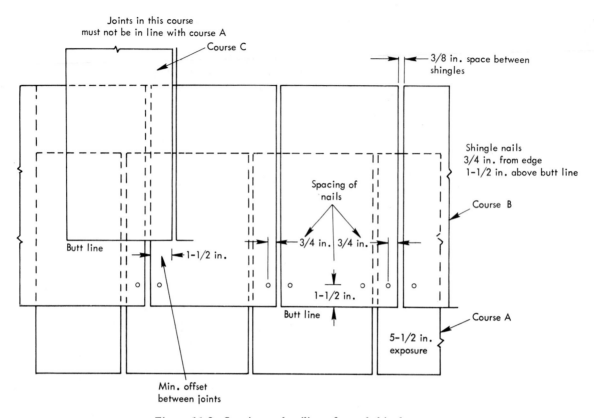

Joints in this course
must not be in line with course A

Course C

3/8 in. space between
shingles

Shingle nails
3/4 in. from edge
1-1/2 in. above butt line

Course B

Spacing of
nails

3/4 in. | 3/4 in.

Butt line

1-1/2 in.

1-1/2 in.

Butt line

Course A

5-1/2 in.
exposure

Min. offset
between joints

Figure 11.3. Spacing and nailing of wood shingles.

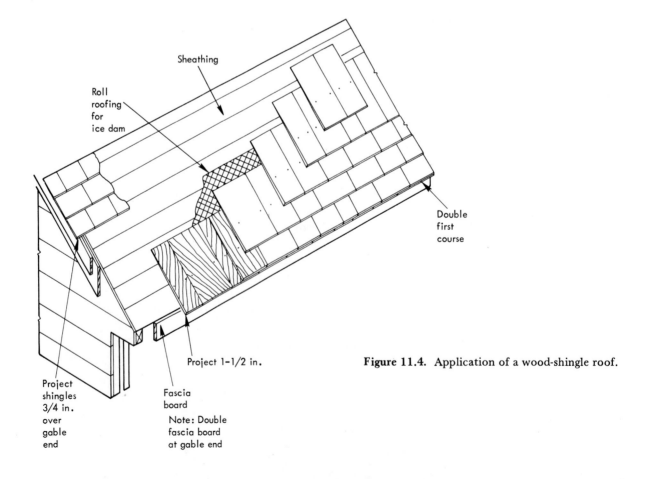

Sheathing

Roll
roofing
for
ice dam

Double
first
course

Project 1-1/2 in.

Figure 11.4. Application of a wood-shingle roof.

Project
shingles
3/4 in.
over
gable
end

Fascia
board

Note: Double
fascia board
at gable end

The laying of shakes is somewhat the same as for shingles but because shakes are thicker and rougher, more precautions must be taken to prevent wind-driven rain or snow from working its way under the courses. Shakes of longer lengths can be exposed more, as shown in Table 11.1. In severe climates it is recommended that not only should a 36-in.-wide strip of heavy roofing felt be placed under the first course (as for shingles), but that an 18-in. felt strip be placed between each course at a distance of twice the weather exposure above the butt line. Example: 5½-in. exposure. Felt placed at 11 in. above the butt line. It is also recommended that in severe climates a fully sheathed roof be used under shakes. Figure 11.4 shows how a wood-shingle roof is laid.

11.2.2 Asphalt Shingles

Asphalt shingles are probably the most widely used roof covering in the country. Properly laid asphalt shingles have a life expectancy of from 15 to 20 years and can be installed to resist high winds and wind-driven rain or snow. Asphalt shingles are available in several weights. The weight most commonly used is the 235-lb square butt strip shingle (the weight is per 100-ft² square). The shingle comes in strips of 12 by 36 in. with 3-in. tab cutouts between the three shingles in a strip. The shingle is available in heavier weights. There are 27 strips in a bundle and three bundles cover 100 ft². The best shingle now comes with adhesive dots placed a little way behind the butt end. These adhesive dots adhere to the shingle above and prevent wind from raising the shingle. Some builders place several dots of asphalt roof cement under the next course as they lay the shingles.

Asphalt shingles are started over a single course of wood shingles which overlap the eave line by 1½ in. The first course is doubled and should overlap the wood shingle by ½ in. The shingles should also overlap the gable

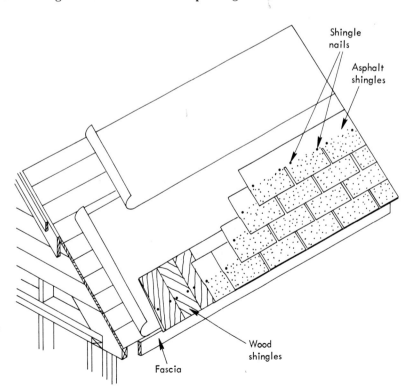

Figure 11.5.
Application of asphalt shingles.

end by $\frac{1}{2}$ in. A 36-in. heavy felt strip is recommended under the first course. Asphalt shingles are usually laid at 5 in. to the weather. A chalk line should be snapped on the shingle strips to guide the worker in laying them in a perfectly straight line. Each strip should be nailed with six 1-in. rust-resistant nails. Figure 11.5 shows a typical asphalt shingle roof.

Asphalt shingles can be used on roofs with a fairly small slope, but in such cases the shingles must be doubled or tripled and there must be an underlayment of 15-lb saturated felt. Table 11.2 shows the necessary coverage for *roofs with small slopes.*

TABLE 11.2. Asphalt Shingle Coverage for Roofs with Small Slopes

	Minimum Roof Slope (ft)	
Underlayment	*Double Coverage*	*Triple Coverage*
None[a]		4 in 12
Single layer	4 in 12	3 in 12
Double layer	2 in 12	2 in 12

[a]No underlayment is required for a roof with a slope greater than 4 in 12.

11.2.3   Built-up Roofing

Built-up roofing is used on flat surfaces. It consists of several layers of saturated roofing paper or felt laid in hot pitch or hot asphalt. Because a flat roof does not allow water to run off rapidly, it must be able to hold water without any leakage. Built-up roofing is designated by the number of years for which a roof can be guaranteed or bonded (i.e., a 10-, 15-, or 20-year roof). The period of guarantee depends on the manner in which the roof is laid and how many layers of felt and pitch are used. A typical architect's specification for a flat built-up roof might read as follows:

> Roofing shall be 20-year bonded type by XYZ Corporation, System ABC, as follows: Apply dry layer of 15-lb asphalt felt nailed to roof deck. Mop 1.2 in. of urethane roof insulation on to felt with steep asphalt. Mop on vapor-bar layer lapped 4 in. on sides, 6 in. on ends. Apply two layers of XYZ 30-lb roofing felt lapped 19 in. with steep asphalt 23 lb per 100 ft^2. Extend all plys up face of cant. Apply gravel surfacing over XYZ specification pitch.

Figure 11.6 shows another version of a roof, which has a suggested service life of 20 years.

11.2.4   Other Roof Coverings

Asbestos shingle is used when a fireproof roof is desired. It can be used within the fire districts of communities where such areas are established. It also has a long useful life and has good insulation value. The shingle is made of asbestos fiber and portland cement and is molded under hydraulic pressure. It comes in natural gray and in several colors. Asbestos shingles

CertainTeed

nailable deck

AVAILABLE NATIONALLY

U.L. CLASS "A"

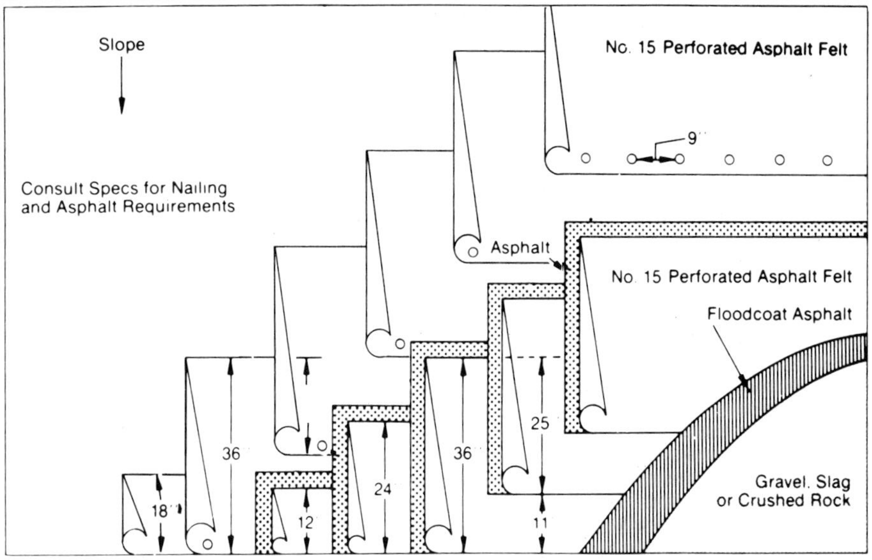

Reference Numbers:	515 GSW
Materials:	No. 15 Perforated Asphalt Felt
Deck Types:	Wood, Plywood, Poured Gypsum, Structural Cement—Wood Fiber, Lightweight Concrete
Slope (incline):	3″ per foot maximum ¼″ per foot minimum
Surface:	Gravel, Slag or Crushed Rock

Materials per 100 sq ft (roof area)

CertainTeed No. 15 Asphalt
Perforated Felt _____ 5 plies
Asphalt (3 moppings) _____ 75 lbs
Flood Coat Asphalt _____ 60 lbs
Surfacing: Gravel _____ 400 lbs
           Slag _____ 300 lbs
           Crushed Rock _____ 450 lbs

Figure 11.6. CertainTeed nailable deck.
(Courtesy of CertainTeed Corporation)

are laid in the same way as wood shingles. They can be cut into the necessary shapes by ordinary carpenters' tools.

Slate shingles are formed by splitting large blocks of natural slate along well-defined cleavage lines. Slate is available in smooth commercial grade or in quarry run, which is rough. It comes in random widths of from 6 to 14 in., in lengths of from 10 to 26 in., and in butt thickness of from 3/16 to ¼ in. for the smooth and from 3/16 to 1 in. for the rough. Copper nails must be used. It is not recommended for any slope under 4 ft in 12 and should be laid over an underlayment of 30-lb felt. The laying of slate roofs requires experienced workers. Falling slates are dangerous. It is heavy (from 700 to 1800 lb per square), it is fragile, and it is expensive, but if it is properly laid, it will last indefinitely.

Clay tile is used in moderate climates on expensive houses. It is available in many shapes, sizes, and weights in the parts of the country where it is most used. Some shapes are French corrugated, Spanish rounded, barrel-curved Mission, and flat shingle. It is heavy (from 800 to 1600 lb per square) and must be laid over 30- to 45-lb felt. Sizes for the flat shingles run from 5- to 8-in. widths; 12-, 15-, and 24-in. lengths; and $\frac{3}{8}$ - to 1-in. butts. It can be used on slopes of over 4 ft in 12 and is fastened with copper nails. Like slate, it is expensive and must be handled by skilled workers, but it will last indefinitely.

Cement tile is used a great deal in warm climates. It can be used on slopes of as small as $2\frac{1}{2}$ in 12. It comes in sizes of $15 \times 8\frac{1}{4}$ and $15\frac{3}{4}$ by $8\frac{3}{4}$. It is set in a bed of cement mortar over an underlayment of 30-lb felt plus 90-lb mineral-surfaced roll roofing.

Fiberglass shingles have a mat of inorganic fiberglass and are therefore more durable and fire-resistant than asphalt shingles, which are backed by an organic felt mat. They are available in three-tab, 12- by 36-in. strips with adhesive strips to keep the courses tightly together.

There are other roofing materials, such as plastics, but they are in an experimental stage and are not yet ready for the commercial market.

11.3   THE FINISHING OF ROOFS

11.3.1   The Purpose of the Finishing

Although the surface of a roof may be completely watertight, it may still leak and leak badly if care is not taken in finishing its perimeter edges, which are its most vulnerable part. Wind-driven rain or sleet may drive through or under a poorly finished edge and completely spoil an otherwise well-laid roof.

11.3.2   Asphalt- and Wood-Shingle Pitched Roofs

Pitched roofs have edges at gable ends, at the lowest starting course and at the gable. Previous sections have described the approved methods of water-proofing the bottom edges and the gable end. Figure 11.4 shows a doubled fascia board along the gable edges. There can also be a fascia board and cornice molding or a fascia board and a special metal strip which is partly under the shingles. The wood or asphalt shingles lap over the edge of the wood or metal edges (at least $\frac{3}{4}$ in.) and complete the waterproofing.

On an asphalt-shingle roof, the ridge is formed by the use of a series of overlapping shingles each 12 x 12 in. The downward slope of the ridge shingle is therefore lapped almost 6 in. on each side over the top course. The ridge shingles can be made by cutting a 12- by 36-in. strip at each tab. They should overlap each other by 6 in. so as to give double coverage (Figure 11.7).

On a wood-shingle roof, the ridge is formed by 6-in.-wide shingles, which are alternately lapped at the edges and blind-nailed and are then lapped longitudinally with no more than 6 or 7 in. exposed (Figure 11.8). To make sure that the ridge is leak-free, it is recommended that heavy roofing felt

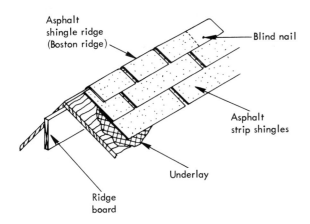

Figure 11.7.
Ridge finish on asphalt shingle roof.

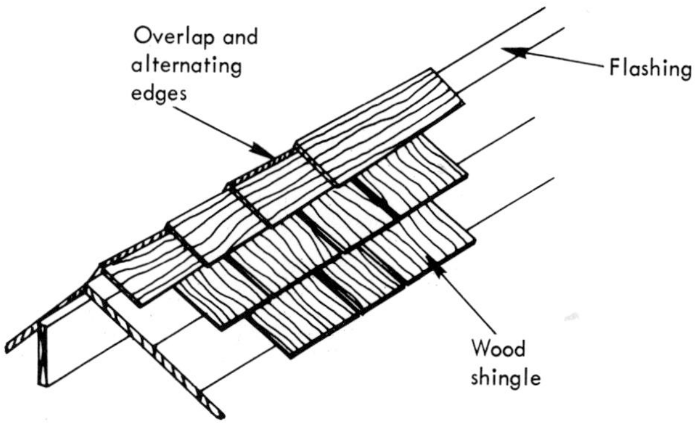

Figure 11.8.
Treatment of ridge on wood shingle roof.

be lapped over the ridge before the ridge covering is applied. There are also ridge coverings made of metal which overlap and, in addition, can be used for ventilation.

11.3.3   Built-up Roofing

A flat built-up roof may end at a roof parapet, at a gravel stop, or, in the case of a porch or other house extension, against a wall. When the roof ends against a parapet wall, a triangular-shaped wood strip is used. This cant strip is placed on the bare roof and against the parapet. The roofing is laid over this cant strip and for several inches up the side of the parapet. A piece of rust-resistant metal is then laid over it, as shown in Figure 11.9.

The same method of base and cap flashing is used when there is a house wall instead of a parapet except that the cap flashing is not bent. This method of finishing is not normally used against a masonry parapet where metal flashing is used, as shown in Figure 11.12.

Figure 11.9.
Use of cant strip at parapet or other wall.

Figure 11.10.
Gravel stop "A," 6-in. strip; "B," 9-in. strip over 6-in. strip

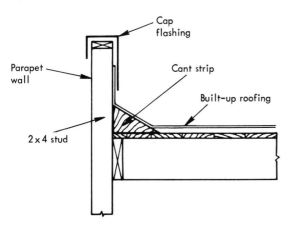

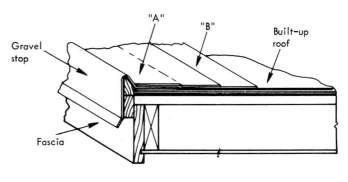

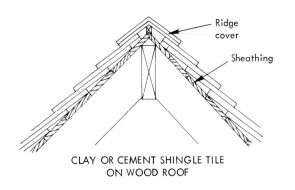

CLAY OR CEMENT SHINGLE TILE
ON WOOD ROOF

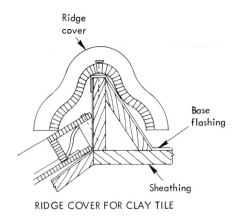

RIDGE COVER FOR CLAY TILE

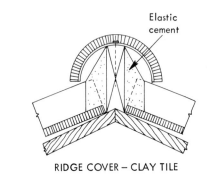

RIDGE COVER – CLAY TILE

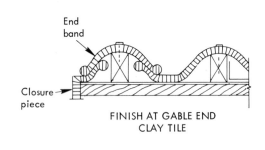

FINISH AT GABLE END
CLAY TILE

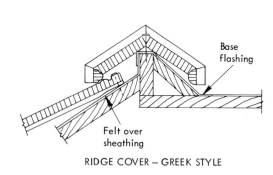

RIDGE COVER – GREEK STYLE

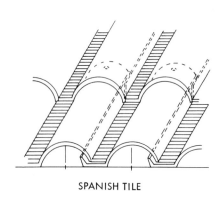

SPANISH TILE

Figure 11.11.
Ridge and gable end covers.

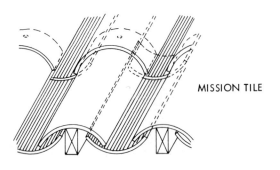

MISSION TILE

188

When a built-up roof ends at a sheer edge, it is finished with a gravel stop. A gravel stop is a strip of rust-resistant metal which is fastened over the first or second layer of roof covering and which projects up above the roof surface and then bends downward over a fascia board as shown in Figure 11.10. The purpose of this strip is to stop gravel roof surfacing from washing over the edge, but it is used at any flat roof edge as a finish for the perimeter edges of the roofing.

11.3.4 Other Roof Coverings

The manufacturers of roof coverings used on peaked roofs, such as asbestos, slate, clay tile, and cement tile, also manufacture shapes for covering the ridges. Figure 11.11 shows how some of these shapes look and how they are applied. The bottom courses are usually finished by double layers of the roofing material, together with the use of heavy roll roofing, and all such products require an underlayment of impregnated roofing felt. The gable ends are finished in the same way as for other peaked roof coverings.

11.4 FLASHING

11.4.1 The Purpose of Flashing

Flashing is the term used for a metal or waterproof fabric that bridges the gap between abutting surfaces in different planes, such as when a roof meets a vertical wall or where peaked roofs meet at a valley. It is also used as a waterproof bridge between unlike materials, such as where siding meets a window or door or when stucco or brick meet wood siding. Without adequate flashing there is no good way to keep wind-driven rain, or moisture from melting snow or ice from entering the open joints that are inevitable when different materials or different surfaces meet.

11.4.2 The Materials of Flashing

The metal flashings must be of rust-resistant materials.

Galvanized sheet metal can be used when the area of construction is away from salt water or corrosive atmospheric conditions. It is recommended that wherever it is used, the sheet steel be 26 gauge or heavier and that the zinc coating (the galvanizing) weigh $1\frac{1}{2}$ ounces per square foot of metal. This gauge of metal and weight of zinc covering will be worth the small difference in price between it and lighter metal and covering.

Aluminum flashing has wide use but, again, it will corrode near salt water or in corrosive atmospheric conditions. The use of aluminum is also not recommended when it comes into contact with stucco or concrete. A minimum thickness of 0.019 in. is recommended.

Copper flashing is one of the most reliable of metals. It can be used anywhere and remains in excellent condition for the life of the building. The use of the various weights of copper for various flashing requirements is given in Table 11.3.

TABLE 11.3. Dimension of Selected Flashing Materials

	Wall Openings	Valleys Gutters	Ridges Hips	Roof Edge	Base Cap
Galvanized steel	26 gauge	24 gauge	24 gauge	26 gauge	26 gauge
Aluminum	0.019 in.	0.019 in.	0.019 in.	0.019 in.	0.019 in.
Copper	10 oz	16 oz	16 oz	16 oz	16 oz
Stainless steel	30 gauge	26 gauge	26 gauge	26 gauge	26 gauge
Terne (painted)	40 lb	40 lb	20 lb	20 lb	20 lb

Stainless steel flashing is often used in high-quality construction. It is noncorrosive and expensive and is not as easy to work as is soft-rolled copper.

Terne plate, which is a thin sheet of steel coated with lead and some tin, is also used as a flashing material. It is expensive and is used only in high-quality construction.

Coated fabric is also used for certain flashing purposes, as at parapet walls and between a roof and a vertical wall. It is made of heavy fabric impregnated with asphalt and is often strengthened by fibers. It is not often used for residential purposes.

All the metals listed above must be fastened with nails of the same metal to prevent corrosive electrolytic action which occurs between different metals when they are in close contact.

Table 11.3 can be used as a guide in determining the proper thicknesses and weights of metal flashings for use in various locations on the building.

11.4.3   The Use of Flashing

Abutting Surfaces on Different Planes

The following illustrations show the use of flashing when two surfaces in different planes meet. Figure 11.12 shows the use of flashing when a flat roof meets a masonry parapet wall. This shows the metal flashing (base flashing) laid over the first course of roofing and then turned up over a cant strip and then covered over by a cap flashing. It also shows a single flashing which can be used over a low parapet. Fabric flashing can also be used in such cases.

Figure 11.12. Flashing: (a) cap and base flashing; (b) cap flashing for low parapets.

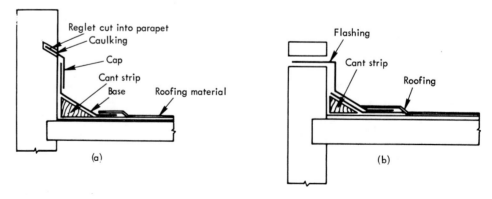

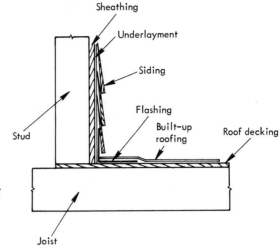

Figure 11.13. Flashing at meeting of flat roof and wood siding.

Figure 11.13 shows how flashing is used when a flat roof meets a vertical wall covered with wood siding. The same type of flashing is used where a sloping roof meets horizontal siding or shingles. The flashing can be cut into pieces which should be at least 8 in. square so that the flashing can extend 4 in. under the roof shingles and 4 in. up the wall. The pieces of flashing must lap each other at least 3 in.

Figure 11.14 shows how the flashing is applied. When a sloping roof meets a brick wall, the flashing procedure is more complicated. There must be a base flashing which extends at least 4 in. under the roof shingles and 4 in. up the wall. This base flashing is covered by a cap flashing, which is let into a reglet cut in the brick wall and caulked. The cap flashing then laps over the base flashing by 4 in., as shown in Figure 11.15.

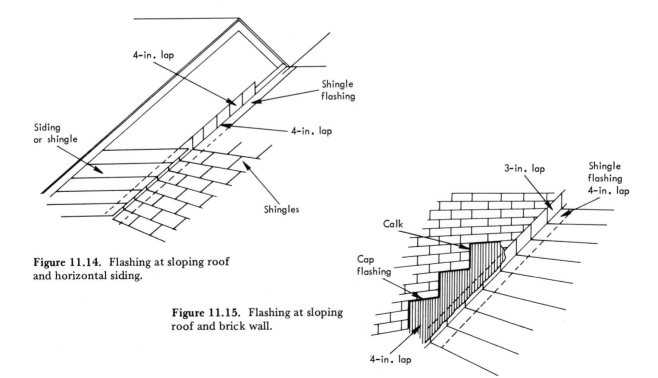

Figure 11.14. Flashing at sloping roof and horizontal siding.

Figure 11.15. Flashing at sloping roof and brick wall.

When two sloping roofs meet at an angle the resultant junction is known as a valley. Valleys are flashed by a continuous strip of metal running from the top junction to the bottom edge of the roof. The flashing must be wide enough to prevent the water from heavy rains or melting snow from finding its way under the shingles. A good rule to follow is to have the valley flashing at least 12 in. wide for slopes of 7 ft in 12 or more, 18 in. wide for slopes down to 4 ft in 12, and 24 in. wide for slopes less than this, although the use of shingles for a slope less than 3 ft in 12 is not recommended unless there is a watertight layer under the shingles. The exposed width of the valley flashing should expand by $\frac{1}{8}$ in. per foot from 4 in. at the top to $4 + \frac{1}{8}$ in. per foot at the bottom. The reason for this widening is, of course, to take care of the increased amount of water at the bottom of the slope. Figure 11.16 shows typical valley flashing. Valley flashing can also be of heavy mineral-surfaced roll roofing.

At Changes in Material

When two different materials meet on an exterior surface, the use of flashing is necessary to bridge the junction. Such junctions occur where brick or stone or stucco join clapboard or shingle or when horizontal and vertical siding meet.

A typical example is shown in Figure 11.17 in the case where a stucco gable end meets clapboard siding. A molding is placed at the top edge of the clapboard or shingle and then a strip of metal is fastened under the sheathing paper and is bent over the molding and then away from it to provide a sharp-edged drip. Figure 11.17 shows how this is accomplished. The flashing must lap the wall at least 4 in. and the roofer must be sure to bend the flashing away from the molding.

The same type of flashing is used when wood siding covers the upper portion of a structure above brick or stone masonry. The flashing starts at least 4 in. up the sheathing behind the siding and ends over the masonry with a sharp drip edge.

Figure 10.11 shows how flashing is used at the base of masonry over a wood structural frame. The flashing on the vertical wall is always fastened behind the sheathing paper.

At Doors, Windows, and Other Wall Openings

The same type of flashing as shown in Figure 11.18 should be used over window and door openings. Figure 11.18 shows a simplified section of a door or window head with the flashing projecting over the top molding. Flashing is also used under the sills of these openings in both wood or masonry wall cladding. Any openings in exterior walls must be flashed the same as for doors and windows—both at the head and at the sill.

Projections through Roof

In residential construction the most important projection through the roof is the chimney. Because of the length of the perimeter of the chimney where it pierces the roof, the chimney flashing must be carefully done.

Figure 11.16. Valley flashing. Note the widening on the downslope.

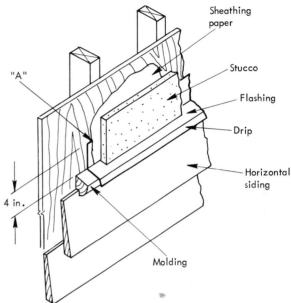

Figure 11.17. Flashing at meeting of stucco and siding. Note that "A" flashing is under sheathing paper.

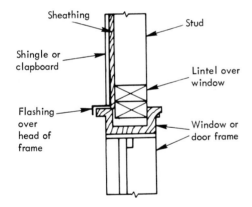

Figure 11.18. Flashing over door or window frame.

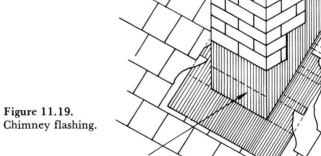

Figure 11.19. Chimney flashing.

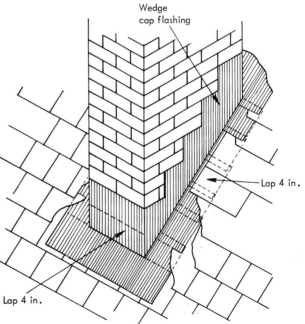

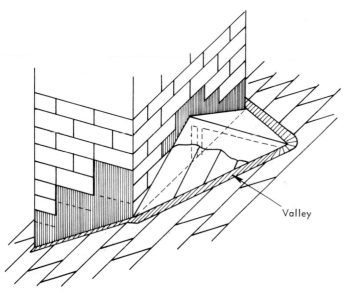

Valley

Figure 11.20. (Left)
Saddle on high side of wide chimney.

Figure 11.21. (Below)
Flashing of chimney at ridge.

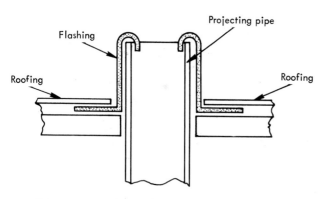

Flashing

Projecting pipe

Roofing

Roofing

Figure 11.22. (Above)
Flashing of a projecting pipe.

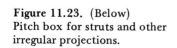

Lap seam soldered

Figure 11.23. (Below)
Pitch box for struts and other
irregular projections.

Figure 11.24. (Below)
Flashing for flagpole, etc.

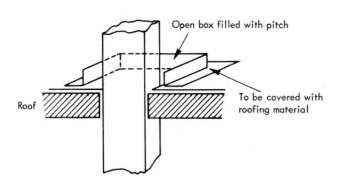

Open box filled with pitch

Roof

To be covered with
roofing material

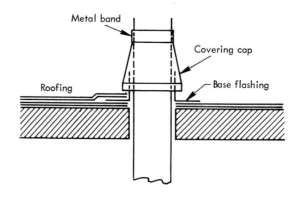

Metal band

Covering cap

Roofing

Base flashing

194

The flashing is used not only between different planes but also between different materials. Figure 11.19 shows chimney flashing. The base flashing should lap under the shingles for at least 4 in. The cap or counterflashing should lap over the base flashing by the same distance. When laying the chimney, the mason should rake out the brick joints in steps as shown, and the counterflashing is inserted into these raked joints and held with lead wedges and caulking. When a chimney on a downslope is particularly wide, it is recommended that a saddle or cricket should be built on its high side to prevent dammed-up rainwater or melting snow from washing over the top of the counterflashing. Figure 11.20 shows how a saddle is built and flashed. The valleys at the sides of the saddle divert most of the water as it pours down from the upper slope. When a chimney comes through the peak of the roof, it is flashed on both slopes, as shown in Figure 11.21.

Roof ventilation or skylights are available with their own integral curbing and flashing, which is covered by shingles or built-up roofing the same as other flashing. Figures 11.22 to 11.24 show the methods of flashing plumbing vent lines, struts, or other projections and a flagpole or antenna pole.

11.5 GUTTERS AND DOWNSPOUTS

Gutters and downspouts are used to collect the water draining from a roof and to guide it away from the house and its foundations. In structures with large roof areas, and in densely populated areas, the water may be led into a storm sewer by means of downspouts located within the structure. In the usual residential construction the gutters and downspouts are located at the edges of the roof and the water is either led into a dry well located away from the structures or falls upon a splash block located at the bottom of the downspout. The splash block is a precast concrete form which serves to distribute the water away from the foundation.

11.5.1   Available Materials and Methods of Installation

Gutters and downspouts are available in galvanized sheet steel, aluminum, and copper. Some steel and aluminum gutters have a baked-on enamel finish. They are now also available in solid vinyl, which is slightly more expensive than aluminum but less expensive than copper. Gutters are also made of wood and are fastened directly to the fascia board at the edge of the roof. This makes them a part of the house structure and presents a better appearance than does a hanging gutter. Wood gutters come in stock sizes of 3 x 4, 4 x 4, 4 x 5, 4 x 6, and 5 x 7.

The following table gives the recommended thicknesses of metal gutters. The "girth" is the width of metal from which the gutter is fabricated.

Girth (in.)	Galvanized Steel (Gauge)	Copper (oz)	Aluminum (in.)
Up to 15	26	16	0.025
16 to 20	24	16	0.032

Larger girths than these are rarely used in ordinary residential construction.

Wood gutters are made of fir, redwood, western red cedar, or cypress. If they are free of knots and are made of heartwood, they will last indefinitely. However, it is recommended that they be given several coats of water-repellent wood preservative.

Metal gutters are available in several sizes and shapes. Figure 11.25 shows the standard shapes. Stock sizes for the ogee shape vary from 2½ in. high by 3 in. wide to 6 by 8 in. and for the half-round from 4 in. wide to 8 in. wide. However, they can be fabricated in any size that is required. The size of the downspouts (leaders), which in turn determine the size of the gutters, is determined by the area of the roof and the probable intensity of the rainfall in that location. In most of the country the recommendation is 1 in.^{2} of downspout per 100 ft^2 of roof. Gutter areas should be the same as the downspout area for gutter runs up to 40 ft. They should be larger for longer runs.

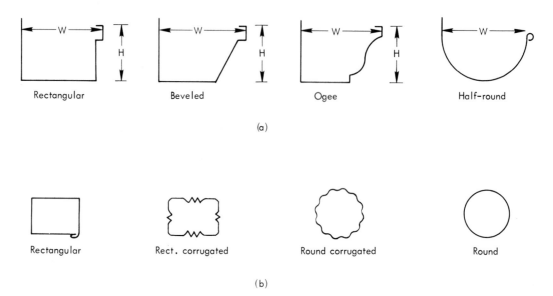

Figure 11.25. Stock shapes of (a) gutters and (b) leaders.

Figure 11.26 shows the various parts of a typical metal gutter and downspout installation and Figures 11.27 and 11.28 show an installation of a wood gutter and a picture of a hanging metal gutter, downspout, and splash block.

As shown, wood gutters are fastened on fascia strips over furring blocks and should be fastened with brass screws. Metal gutters are hung on metal hangers of like metal which are fastened to the roof under the shingles or other roof covering. Hangers should be not more than 30 in. apart. Gutters should slope toward the downspouts. Leaders (downspouts) are fastened to the wall with straps or spikes.

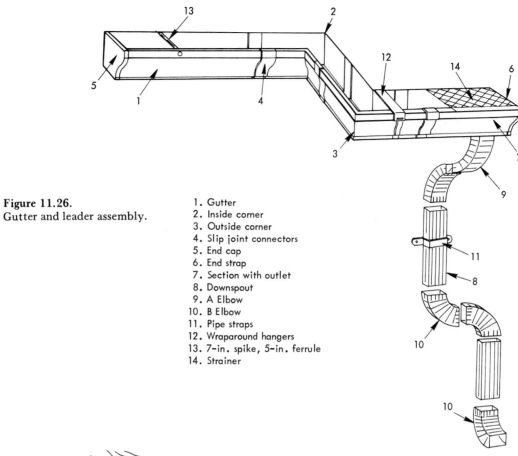

Figure 11.26.
Gutter and leader assembly.

1. Gutter
2. Inside corner
3. Outside corner
4. Slip joint connectors
5. End cap
6. End strap
7. Section with outlet
8. Downspout
9. A Elbow
10. B Elbow
11. Pipe straps
12. Wraparound hangers
13. 7-in. spike, 5-in. ferrule
14. Strainer

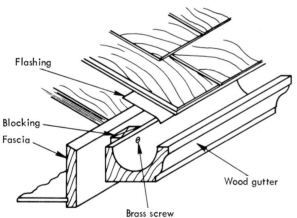

Figure 11.27.
Wood gutter installation.

Flashing

Blocking

Fascia

Brass screw

θ

Wood gutter

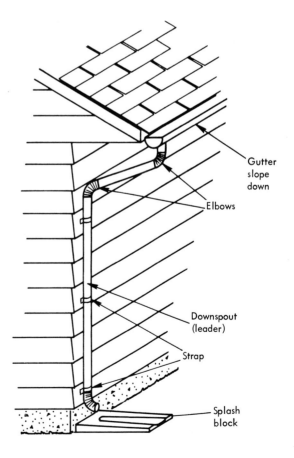

Figure 11.28.
Hanging metal gutter assembly.

Gutter slope down

Elbows

Downspout (leader)

Strap

Splash block

CHAPTER TWELVE

interior walls, ceilings, and floors

12.1 DESIRED CHARACTERISTICS OF INTERIOR WALL AND CEILING FINISHES

The interior finish of a building must have certain qualities that will enable it to function best in the particular part of the structure where it will be placed. It must be durable; of the proper thickness; either prefinished or serve as a base for paint, wallpaper, or other finishes; and in certain areas it must be moisture- and fire-resistant. In choosing the type of finish to use, the contractor should become familiar with the finishes that are available, the best use for each, the method of its installation, and the comparative cost.

12.2 AVAILABLE MATERIALS AND THEIR BEST USE

The materials most frequently used for interior finish are described below. They meet a wide variety of requirements—economy, quality, appearance, and desirability.

12.2.1 Gypsum Wallboard

This material is used as a finish for partitions, exterior walls, and ceilings. It is called drywall—as differentiated from plaster walls. Gypsum wallboard or plasterboard is manufactured in sheets of various sizes and thicknesses. It consists of gypsum mixed with fiber and faced on both sides and the edges with paper. It can be obtained in its usual standard size of 4 ft by 8 ft, but the length can go to 16 ft when it is to be used for horizontal application. It also comes in standard thicknesses of $\frac{1}{4}$, $\frac{3}{8}$, $\frac{1}{2}$, $\frac{5}{8}$, and 1 in.

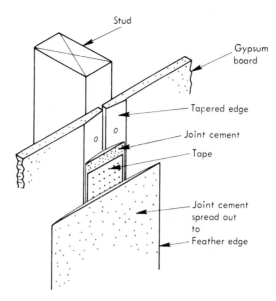

Figure 12.1. Applying feather
· edge to gypsum board.

The sheets can be obtained with foil backing, which is used as reflective insulation and a vapor barrier. They are also available with vinyl or other prefinished surfaces, which eliminate the necessity of painting or wallpapering. The usual sheet comes with a tapered edge so that the surface of the finished joint will not protrude above the surface of the sheet. Figure 12.1 shows how a feather edge is applied (see also Figure 12.5).

Methods of Application

Table 12.1 gives the recommended thickness and weight of gypsum board for various stud, joist, or rafter spacing. The ¼-in. thickness is not recommended for new construction but rather as a covering over existing walls.

TABLE 12.1. Recommended Gypsum Board Dimensions for Various Spacing Applications

Thickness (in.)	Weight per square foot	Method of Application	Maximum Spacing of Framing (in.)
		Walls	
⅜	1.5	Vertical	16
½	2.0	Vertical	24
⅝	2.5	Vertical	24
⅜		Horizontal	16
½		Horizontal	24
⅝		Horizontal	24
		Ceilings	
⅜		Right angle to joists	16
½		Right angle to joists	24
		Parallel to joists	16
⅝		Right angle to joists	24
		Parallel to joists	16

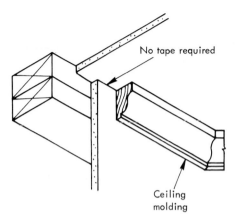

Figure 12.2. Butting of ceiling and wall with ceiling molding.

The ½-in. thickness is the one most often used in ordinary construction. It can be applied vertically or horizontally. The horizontal application is useful when full sheets can be used without cutting, and in longer than the usual 8-ft lengths. Sheets of the longer length must be handled carefully to prevent bending and cracking.

When studs or joists are spaced 16 or 24 in. on center, a 48-in.-wide sheet will cover two or three structural members. The sheets should be nailed on 5- to 7-in. centers on ceilings and on 6- to 8-in. centers on walls. The sheets should not butt too closely, to leave some space for the joint finishing and taping material to key in.

When installing gypsum board, the ceilings should be completed before the walls are started. Figure 12.2 shows how ceilings and walls can be butted without taping. The builder must be very careful to see that the joists and studs are correctly spaced and are exactly parallel. If this is not done, the sheets may not meet exactly in the center of the joists or studs. As drywall is primarily used because of its ease of construction, uneven framing defeats the whole purpose. Figures 12.3 and 12.4 show gypsum board in vertical and horizontal applications.

The builder or owner-builder who uses gypsum board for interior finish for walls and ceilings should be aware that the taping and finishing of the

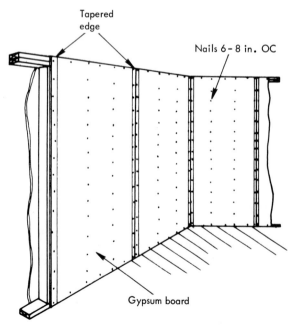

Figure 12.3.
Gypsum wall board applied vertically.

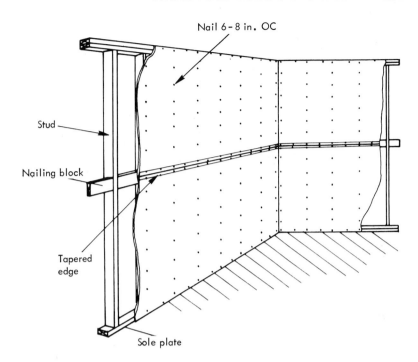

Figure 12.4. Gypsum wall board applied horizontally.

joints between the sheets is a highly skilled operation. The finishing of the joints is accomplished in three separate operations which are usually spaced at least 24 hours apart, depending on temperature and humidity.

Figure 12.5 shows the separate operations that are necessary to ensure a perfect joint. A wide spackling knife should be used. The builder must also see to it that the gypsum board is stored in a dry place and kept dry. The wood framing members to which the sheets will be nailed should be allowed to dry enough so that they will not shrink after the board has been applied and either pop the nails or crack the sheets. It can be a very expensive matter if the interior finish has to be renailed and retaped. In most of the country the wood should attain an 8% moisture content before the gypsum sheets are installed. In humid states an 11% moisture is recommended, and in the arid parts of the country the moisture should not exceed 6%. A moisture meter can be used and it is well worthwhile to use it.

12.2.2 Lath and Plaster

The use of lath and plaster as an interior wall and ceiling finish has been steadily diminishing as the cost of construction has been increasing. However, it is still used extensively in high-quality construction. It is sturdier than wallboard and is more resistant to fire and sound and heat transmission. It is more expensive than wallboard and its greatest disadvantage, especially in speculative construction, is that it takes several weeks to dry, and during this drying time no finished trim or floors can be installed. The plaster must also be thoroughly dry before any paint (except water-based latex) or wallpaper can be applied. This delay is the basic reason for its diminishing use. It is strongly recommended, however, for "tailor-made" houses that are built for a specific owner.

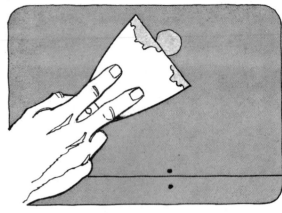

1. Use joint compound to fill space between tapered edges.

2. Spot nail heads with first coat of compound.

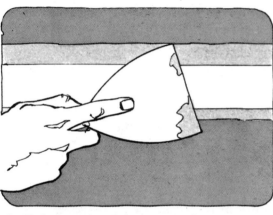

3. Embed tape into joint compound.

4. Smooth joint compound around and over tape and level surface.

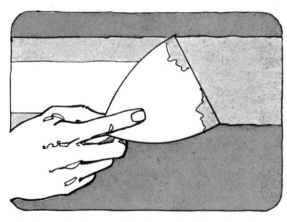

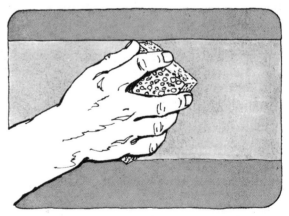

5. First finishing coat (after 24 hours). Apply thinly and feather out 3 or 4 in. on each side.

6. Second finishing coat (again after at least 24 hours). Spread thinly and feather out 6 to 7 in. on each side. Finish nail holes.

Figure 12.5. Joining Gypsum Board Sheets.

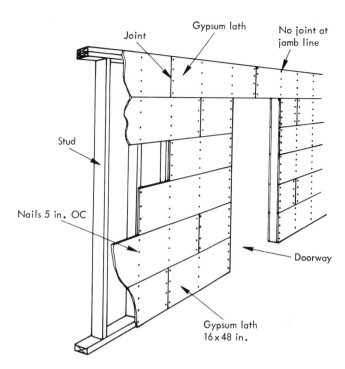

Figure 12.6. Application of gypsum lath.

The Materials and Application

The base for a plaster finish can consist of gypsum lath, which is a paper-faced gypsum-filled board. It comes in two thicknesses, ⅜ and ½ in., and in a standard size of 16 by 48 in. The board also comes with an insulating foil backing and with either a plain or a perforated surface, the latter being used when a particularly strong bond is required. In either case the adhesive of the plaster to the lath is very good. Experiments have shown that it requires a force of over 800 lb per square foot to separate the plaster from the gypsum lath. The gypsum board is nailed to the studs in a horizontal position and with broken joints, as shown in Figure 12.6. To avoid the cracking that occurs as structural wood members shrink, it is good practice to nail metal lath strips over the corner of doorways and where walls meet ceilings, as shown in Figure 12.7. Where there are exposed plaster corners which are subject to breakage, it is recommended that the corners be protected with a metal corner bead, as shown in Figure 12.8.

Figure 12.7. Metal lath reinforcement: (a) over doorway; (b) at corners.

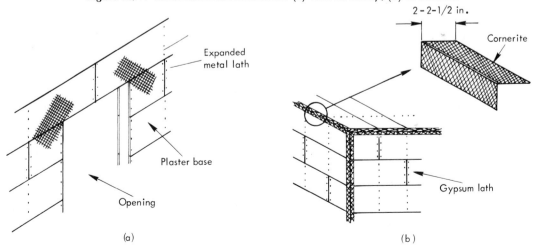

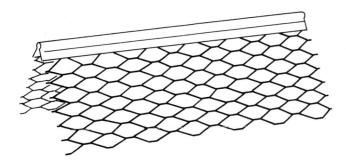

Figure 12.8. Metal corner bead.

Another plaster base consists of insulating board that is made of compressed fiber faced with paper. It has more insulating value than gypsum lath and comes in the same standard size of 16 by 48 in., although other sizes, such as 18 by 48 in. and 24 by 48 in., can be obtained. The standard thickness is ½ or 1 in. This board can be obtained with a metal lath attachment along its long edge. This provides excellent resistance to cracking.

Metal lath is the most versatile of the plaster bases. It offers the best key for the plaster and it adds to the fire resistance. It can be cut and shaped to fit into oddly shaped spaces and it is resistant to cracking. Even when gypsum lath is used, it is desirable to use metal lath in certain strategic places, as indicated earlier in this section.

Figure 12.9 shows two examples of metal lath and an illustration of how it can be used to fur out a bathroom pipe space. Metal lath is specially recommended for use in damp places. It is also used as a base for cement plaster in fire-rated walls, such as the wall between a garage and house or the walls of a furnace room. Metal lath can be nailed directly to wood studs or wood ceiling joists, or it can be wired or clipped to metal studs or hung from a ceiling by suspended metal channels to which the lath is wired.

Plaster finish is applied in two or three coats. The usual thickness of plaster is about ½ in. over gypsum lath and slightly more over metal lath. Three-coat plaster consists of scratch, brown, and white. When gypsum or other board lath is used, the scratch and brown coats are applied in one

Figure 12.9. (a) Typical metal lath. (b) Typical metal rib lath. (c) section of heavy channel for bathroom, etc., wall.

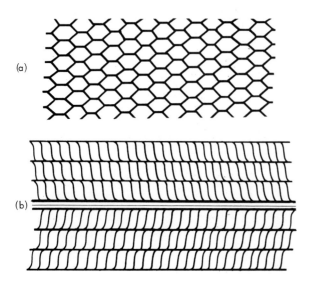

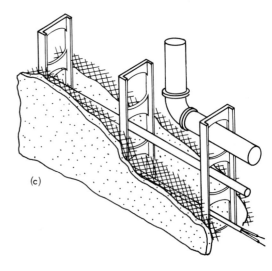

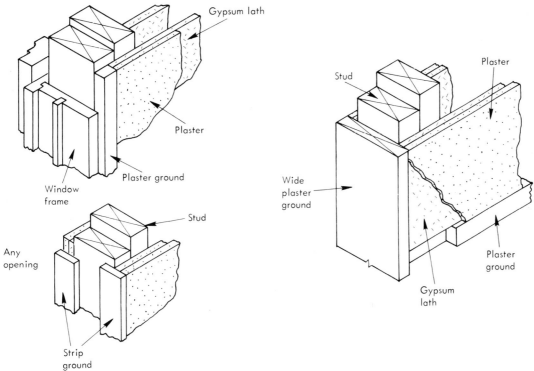

Figure 12.10.
Examples of plaster grounds.

operation, known as "doubling back." The scratch coat is first applied and then before it sets it is scratched or roughened and a second coat is applied to bring the plaster to just below the face of the plaster grounds. Plaster grounds or screeds are strips of smooth wood that are used to define the thickness of the plaster and to help the worker to keep it smooth and level (Figure 12.10). The usual grounds are ⅞ in. thick by 1 in. wide and are used at the floor and at door and window openings. The final coat of plaster is usually white lime putty made by mixing hydrated lime with water and adding gauging plaster. This dries to a hard, smooth coat.

Plaster over metal lath should be applied in three coats, and each coat should be completely dry before the next coat is applied. Plastering is a skilled trade. To produce a hard, smooth-finish white coat or even a sand finish or other rough finish requires a good deal of experience, not only in attaining a smooth and level surface but in the mixing of the materials.

12.2.3 Plywood, Hardboard, and Fiberboard

These materials usually come in 4- by 8-ft sheets but can be longer. The finished surface comes in a wide variety of colors and finishes. It can look like wood paneling in imitation of a number of different woods. It can look like brick or stone, or can come in patterns to look like wallpaper. The backing for these finishes is plywood, Masonite or other hardboard, and fiberboard.

Plywood and hardboard should be at least ¼ in. thick when used over studs that are on 16-in. centers and ⅜ in. thick when studs are on 24-in.

centers. Fiberboard can be used on ceilings as well as on walls. Because it is less sturdy than plywood or hardboard, the thickness should be ½ in. on 16-in.-on-center (OC) studs and ¾ in. on 24-in. OC studs. Fiberboard has more insulating value than either plywood or hardboard. All of these materials can be glued to the studs, in which case less nails can be used and there is less chance of marring the surface.

Fiberboard, fiberglass, and fissured gypsum board are used for acoustic ceilings or for any decorative ceiling. The material comes in 12- by 12-in. squares and can be obtained in other sizes (12 x 24) (24 x 48). It can be installed by glueing it to plaster or drywall or by nailing it to furring strips that are placed at right angles to the joists, as shown in Figure 12.11, or the tiles can be fitted into a metal grid that is suspended from the ceiling, as shown in Figure 12.12.

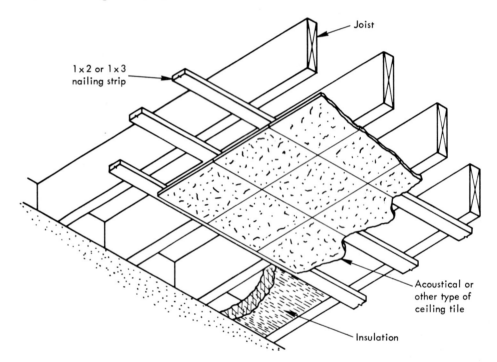

Figure 12.11. Ceiling tile nailed to furring strips.

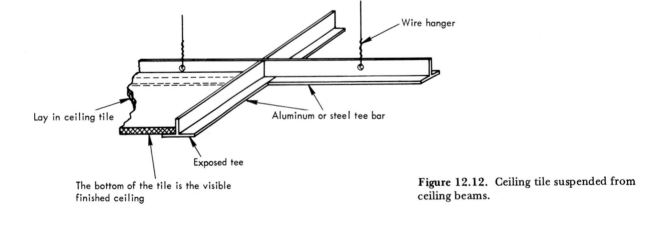

The bottom of the tile is the visible finished ceiling

Figure 12.12. Ceiling tile suspended from ceiling beams.

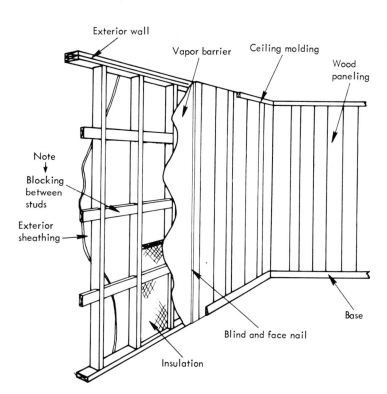

Figure 12.13.
Typical application of wood paneling.

12.2.4 Wood Paneling

Wood paneling is often used as an interior finish. It is decorative and gives the owner a feeling of high quality. It is also quite sturdy and resistant to abuse. Wood paneling must be thoroughly dry. It is suggested that it be kept in a dry, heated room before it is applied. A good average moisture content is 8%. There are many varieties of natural wood paneling available to meet individual tastes. A typical material dealer's stock may include the following:

> No. 2 Idaho knotty pine: 1 x 6, 1 x 8, 1 x 10; V-joint
> Clear-heart vertical-grain redwood: 1 x 6, 1 x 8, 1 x 10; V-joint
> Western red cedar—knotty: 1 x 6, 1 x 8, 1 x 10; V-joint
> Knotty pine—rough sawn: 1 x 8, 1 x 10 shiplap, 1 x 12; square
> West coast hemlock: 1 x 6, 1 x 8, 1 x 10; V-joint or rounded

There are also wormy chestnut, pecky cypress, and others. Wood paneling is usually applied vertically, although the shiplap variety can be used horizontally. The thickness of the wood should be at least ⅜ in. for 16-in. OC studs and ⅝ in. for 24-in. OC studs, and the space between studs must be blocked at 24-in. OC for vertical application. Because of the possibility of shrinkage, the builder or owner is advised to keep most of the boards at the 8-in. width, although some 10-in. widths can be used in a pattern of random width.

Figure 12.13 shows a typical application of wood paneling. The builder must be careful to be sure that the starting board is exactly vertical. When the paneling is on an exterior wall, it is advisable to use a plastic-sheet vapor barrier directly fastened to the studs. The exterior wall insulation need not then have such a barrier.

Although natural wood paneling is much more expensive than most other interior finishes, it is used in selected places in high-quality houses by owners and by speculative builders.

12.3   FLOOR COVERINGS AND FINISHES

12.3.1   The Available Materials and Their Best Use

The material of the finish flooring in residential construction and the method of its installation is determined by its appearance, its wearability, its cost, and its comfort, among other reasons.

Probably the most commonly used floor finish is wood-strip flooring. This is available in soft and hard woods and in a variety of sizes. The soft woods do not wear as well as the hard woods and are more subject to scuffing and abrasion. They are recommended for use in low-traffic areas. The edge grain variety of this wood is the better one for wearing quality. The softwoods that are available are Douglas fir, western larch, eastern hemlock, and southern pine. Thicknesses (in inches) run as follows: $5/16$, $7/16$, $9/16$, $25/32$, $1\frac{1}{16}$, and $1\frac{5}{16}$; face widths are $1\frac{1}{2}$, $2\frac{3}{8}$, $3\frac{1}{4}$, $4\frac{1}{4}$, and $5\frac{3}{16}$. The most popular size is $25/32$ by $2\frac{3}{8}$. At this width the builder should allow 30% for waste (i.e., 30% more flooring than the actual area to be covered). Softwoods come as T&G, shiplap, and in squares held by splines.

The hardwoods that are available are oak, beech, birch, pecan, and northern hard maple. The most extensively used of these is oak-strip flooring $25/32$ by $2\frac{1}{4}$. This is available in T&G at both the sides and end, matched as shown in Figure 12.14. It makes a good hard-wearing floor and has a nice appearance.

Beech, birch, and maple are available in standard thicknesses of $25/32$, $33/32$, $41/32$, and $53/32$ in. and in face widths of $1\frac{1}{2}$, 2, $2\frac{1}{4}$ and $3\frac{1}{4}$ in. These make very handsome and hard-wearing floors.

Both soft- and hardwood flooring is also available in squares which are made up of strips bound together by a metal spline or by adhesives. These blocks come in a number of sizes and make a very pleasing floor pattern (parquet). Some of the sizes available are $7\frac{1}{2}$ by $7\frac{1}{2}$, 8 by 8, 9 by 9, and 10 by 10 in., all $25/32$ in. thick.

There is also cork flooring; particle-board flooring, which is made of wood fibers and resin under high compression; and flooring squares made of prefinished plywood. Slate flooring is often used in entrance halls in high-quality construction. Ceramic tile is often used on bathroom floors and quarry tile is used on kitchen floors.

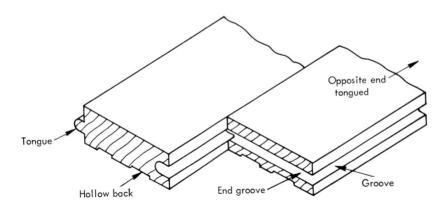

Figure 12.14. Tongue-and-groove and end-matched strip flooring.

Resilient flooring is very often used in houses built on a flat slab and is almost always used in kitchens and bathrooms.

There are many qualities of resilient flooring—from asphalt tile through vinyl asbestos to pure vinyl. There is also rubber, vinyl, and linoleum, which is laid in sheets. The usual resilient flooring is available in squares 9 by 9 or 12 by 12 in. and ⅛ in. thick. Linoleum, rubber, and sheet vinyl are available in rolls that are usually 6 ft wide. A table showing the varying qualities of resilient tile would show that a pure vinyl tile is foremost in durability, resistance to damage, resistance to grease and staining, and is most quiet. This is followed by sheet vinyl, linoleum, vinyl asbestos, rubber, and finally asphalt tile, which is the cheapest and of the least quality.

12.3.2 Methods of Application

Wood Floors

Wood-strip floor is laid over a board subfloor and should be at right angles to the joists. It is blind-nailed, usually with a cut nail that is driven at a 45° angle and then set as shown in Figure 12.15. When it is end-matched, the joints should be staggered. The builder should mark the joist locations with a chalk line and drive as many nails as possible through the subflooring and into the joists. This will help to eliminate squeaking floors, which are caused by inadequate fastening of the floor. The starting strip of the flooring should be laid at a distance of at least ½ in. from the walls to eliminate possible buckling if the floor should swell. Figure 12.16 shows how this is done.

Figure 12.15.
Wood-strip flooring.

Figure 12.16.
How to start a wood-strip floor.

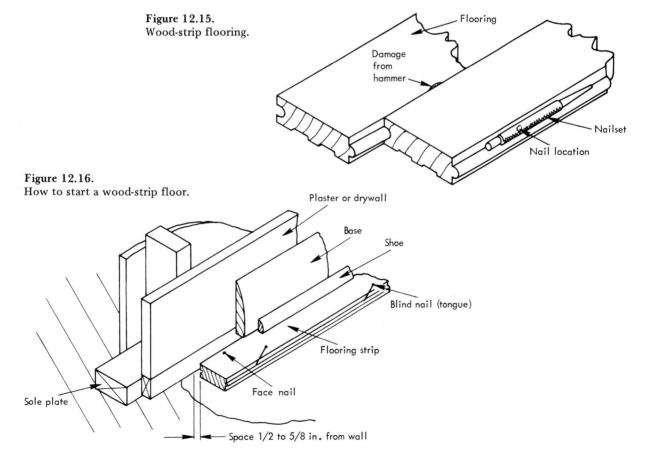

It is most important that wood flooring be kept absolutely dry from the time of its delivery to the time of its installation. It should not be delivered until the house is weathertight and should be kept in a warm, dry place. Wood flooring is not installed until most of the trim has been set. It is best to wait until the joists and underfloor have had a chance to dry to near their final moisture content (8% except as previously mentioned) so as to avoid open cracks caused by shrinkage. After laying, wood or other finish flooring should be covered with heavy building paper.

Wood-strip floor is also used over concrete slabs. It is hoped that the builder has laid a polyethylene or other vapor barrier under the concrete slab. If not, then the top of the slab is covered with a waterproof troweled-on mastic into which 1- by 4-in. treated furring strips are laid on 16-in. centers and anchored to the concrete. A polyethylene sheet is laid over these strips and then the flooring is laid at right angles and nailed to these strips. In high-quality construction, another 1- by 4-in. furring strip is nailed over the first strip and the vapor barrier sheet; then the finish strip floor is nailed to this.

Wood parquet or particle-board flooring is laid in mastic over either a concrete slab or over an underfloor, which is usually of plywood. Parquet flooring is made in squares consisting of small strips with opposing grain held together by metal splines or by a membrane which is removed when the flooring is set. It is best to set this type of flooring in a pattern to form a checkerboard effect. This is a pleasing design and tends to equalize shrinkage and swelling. This flooring must be handled in the same way as strip flooring (i.e., kept warm and dry; Figure 12.17).

Resilient Flooring

Of the various resilient tiles, only asphalt tile can be laid directly on a concrete slab, which must be thoroughly dry. The concrete should be specially leveled and smoothed to avoid dips or other noticeable imperfections, and it should have a vapor barrier under it. The other floors, such as vinyl asbestos, vinyl, and so on, should be laid over an underlayment such as plywood or particle board which is laid over furring strips. Vinyl asbestos is sometimes laid directly on a slab, but this is not recommended.

In frame construction, resilient floors should not be laid on a board underfloor. Smooth-faced plywood or particle board should be laid over the underfloor and the resilient tile is laid over this in a bed of adhesive which is spread with a notched trowel. Many builders simply use ¾-in. plywood directly over the joists as underlayment. In laying any resilient tile, care should be taken that no joints between the tile coincide with joints in the underlayment.

Ceramic or Other Tile Floors

Ceramic or other tile floors can be applied in two ways. In bathrooms or other areas subject to excessive moisture, it is recommended that the tile be laid in a cement grout over a concrete bed as shown in Figure 12.18. Mosaic ceramic tile comes in sheets 1 by 2 ft or 2 by 2 ft and is paper-backed. It is laid face down in the grout and the backing is removed. The concrete bed should be at least 1½ in. thick and reinforced with a light

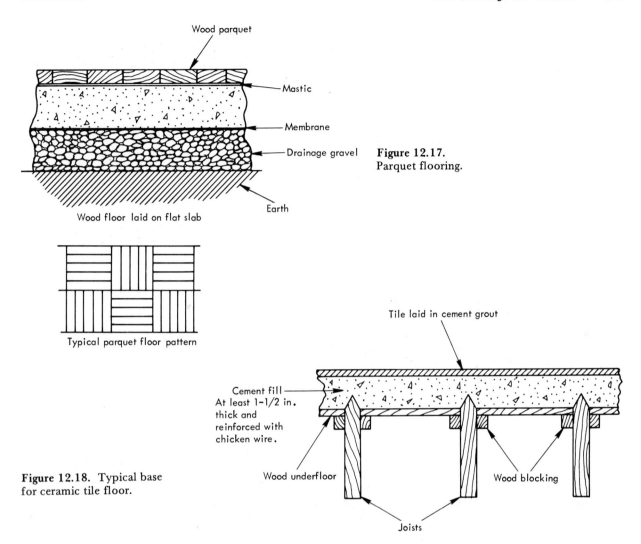

Figure 12.17.
Parquet flooring.

Figure 12.18. Typical base
for ceramic tile floor.

mesh. The joists under should be chamfered as shown. Quarry tile and slate can, of course, be laid directly on a concrete slab. With a wood-frame floor, a ¾-in.-thick waterproofed plywood underlayment is recommended. The tile should be laid in a full bed of adhesive and allowed to partly set before the joints are tooled.

12.3.3 Carpet

It has become very popular for builders of speculative houses to install wall-to-wall carpeting in every room in the house. The trouble in most cases is that the carpet is laid directly over an underfloor, usually plywood, and as a carpet does not wear as long as other floors, it must be replaced at considerable cost or, if the occupant does not want carpet, he or she is faced with the cost of a new finish floor. To save future expense and for re-sale value, it is advisable to have a finish floor under the carpet. Carpet comes in many qualities and in many patterns. However, wool carpeting holds its own in competition with most of the new synthetics.

CHAPTER THIRTEEN

thermal insulation:
the conservation of energy

13.1   THE PURPOSE OF THIS CHAPTER

At the time of this writing and for the foreseeable future no book on building construction can be complete without a discussion of the cost and use of energy and how savings can be made by the proper use of insulation. The problem is the same whether the bulk of the energy is used for heating in the colder climates, for air conditioning in the warmer climates, or for a combination.

Basically, insulation is used to help maintain the required inside temperature of a structure with the minimum use of energy: in other words—maximum comfort at minimum cost. The insulation must therefore serve to retard as much as possible the transmission of heat through any surface. The purpose of this chapter is to guide the builder in the choice of insulation to use, where to use it to the greatest advantage, and the choice of the optimum efficiency of the insulation to best serve the purpose.

As a final word about thermal insulation, there is no part of the country in which a certain minimum of such insulation is not beneficial to either help the air-conditioned house by shielding it from outside heat or helping the heated house to retard the escape of heat to the outside.

13.2   DEFINITION OF TERMS

To avoid overinsulating, which is costly at the present cost of insulating material, or underinsulating, which will cause a waste of energy, the builder should be familiar with the terms that are used in calculating how much insulation should be used and in describing the value of the material.

Btu (British Thermal Unit)

The heat required to raise the temperature of 1 pound of water by 1 degree Fahrenheit.

U

This is expressed as a number. It tells the number of Btu per hour that is transmitted by 1 square foot of surface per 1 degree Fahrenheit in temperature between two spaces. For instance:

Outside temperature 20°F

Inside temperature 68°F

Differential 48°F

A table of *U* values will show that the walls of an uninsulated house with interior walls of ½-in. gypsum board over studs and exterior walls of bevel siding over plywood sheathing has a *U* value of 0.26. This means that every square foot of wall surface will transmit 48 x 0.26 = 12.5 Btu/hr. An example of what this means in terms of dollars will be shown in Section 13.6.

R

The term that is used in describing the value of insulation is *R*, which is the reciprocal of *U*. All insulation is required to have its *R* value stamped or otherwise affixed to it. The higher the *R* value, the better the insulation. As an example, a *U* factor of 0.26 is very poor. The *R* value in this case is 1/0.26, or 4. If the *U* factor goes down to 0.10, for instance, the *R* value is 1/0.10, or 10.

K

Some manuals use this factor. It is a measure of the transmission of heat (Btu) through 1 square foot of material per 1 degree Fahrenheit of difference in temperature per *1 inch of thickness* as compared to *U*, which is for an entire assemblage, such as an exterior wall or a roof.

Vapor Barrier

All insulation between surfaces of different temperatures is subject to moisture caused by condensation that occurs when warm air drops its moisture on encountering a cold surface. Insulation loses a great deal, if not all of its value when it is moist. It is also subject to mildew and rot and by transmitting its moisture through a wall surface it can cause paint peeling and plaster damage. To avoid this, all insulation should be protected by a vapor barrier, which prevents the passage of moisture between the warm surface and the insulation. Vapor barriers can be of metal foil, plastic sheeting, or moisture resistant paper. The vapor barrier should always face the warm surface.

13.3   INSULATING MATERIALS AND THEIR EFFICIENCY

Insulating materials come in various shapes and forms and in efficiency. The efficiency of the material can be known immediately by its stated *R* value (Table 13.1).

TABLE 13.1. Insulation Uses

Material	R per Inch Thickness	Where Best Used
Flexible—batts or blankets		
Glass fiber—mineral wool	3.00–3.40	Rafters, crawl spaces, ducts
Cellulose fiber—organic	3.20–4.00	Attic floors, undersides of floors, open sidewalls before finish
Loose fill		
Glass fiber—mineral wool	2.80–3.40	Inside finished sidewalls, finished
Cellulose—organic	3.50–3.70	or unfinished attic floors
Expanded vermiculite	2.13	
Rigid board		
Polystyrene, extruded	5.26	Floor slab perimeters, basement
Preformed urethane	5.80–6.25	walls, any new construction,
Glass fiberboard	4.00	under finish roofing
Molded polystyrene	3.57	
Foamed in place		
Expanded urethane—sprayed	6.25	Finished sidewalls, finished or unfinished attics
Reflective—two sides	Total R = 4.3	Sidewalls

If organic material is used, the builder must be sure that it has been treated to make it resistant to fire, decay, insects, and vermin.

For quick reference, Table 13.2 may be used. This gives the total R value for the most commonly used materials and thicknesses. The last two thicknesses of material in Table 13.2 are for use only in very severe climates.

TABLE 13.2. R Values for Insulating Materials

R Value	Batts or Blankets (in.)		Loose Fill (in.)		
	Glass Fiber	Rock Wool	Glass Fiber	Rock Wool	Cellulose
11	$3\frac{1}{2}$–4	3	5	4	3
19	6–$6\frac{1}{2}$	$5\frac{1}{4}$	8–9	6–7	5
22	$6\frac{1}{2}$	6	10	7–8	6
30	$9\frac{1}{2}$–$10\frac{1}{2}$	9	13–14	10–11	8
38	12–13	$10\frac{1}{2}$	17–18	13–14	10–11

13.4   WHERE AND HOW TO INSTALL INSULATION

A general rule for the installation of insulation is to place it on every surface that separates heated from unheated spaces so as to minimize heat loss from one to the other. Figure 13.1 is an illustration of where insulation should be installed.

Where to Insulate
1. Ceilings with cold spaces above
2. Rafters and "knee" walls of a finished attic
3. Exterior walls; walls between heated and unheated spaces; dormer walls
4. Floors over outside or unheated spaces
5. Perimeter of a concrete floor slab close to grade level
6. Walls of finished or heated basement
7. Top of foundation or basement wall.

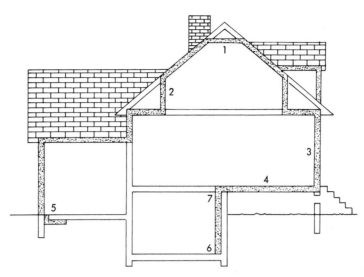

Figure 13.1. Where to insulate a house.

13.4.1   Basements and Crawl Spaces

In houses with unheated crawl spaces, insulation should be placed between the floor joists or around the wall perimeter and it is recommended that a ground cover of roofing paper or polyethylene sheet be placed over the earth of the crawlway to eliminate excessive moisture.

The insulation between the joists is usually of faced batt insulation—that is, mineral or fiberglass wool faced on both sides by treated paper. The vapor barrier should be faced up toward the heated area.

Figure 13.2. Insulation of basement or crawlway ceiling.

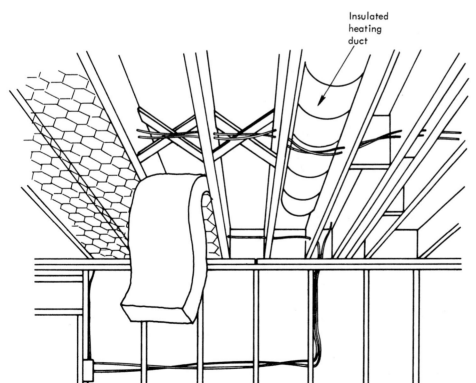

Figure 13.2 shows how the insulation can be installed over light wire mesh. Insulation can also be applied against basement walls by the use of furring strips nailed to the concrete or block and the use of blankets between the strips. Rigid board can also be used as insulation. It can be spot-glued to the underside of the floor between the joists and can also be used against the foundation wall by nailing to furring strips.

13.4.2 Walls and Openings

Exterior walls are to a great extent insulated with flexible batt insulation which is faced on both sides, one side of which is a vapor barrier. In wood-framed houses the blanket is stapled between the studs with the vapor barrier facing in. Sometimes pressed-fit or friction-type insulation is used. This does not have to be stapled, but it must be covered on the inside with a sheet of polyethylene which is stapled to the studs and to the top and bottom plates (Figure 13.3).

Masonry exterior walls should be insulated by fastening furring strips to the walls and then using foil-backed wallboard as reflective insulation. Reflective insulation will only be effective if there is a 1-in. air space in

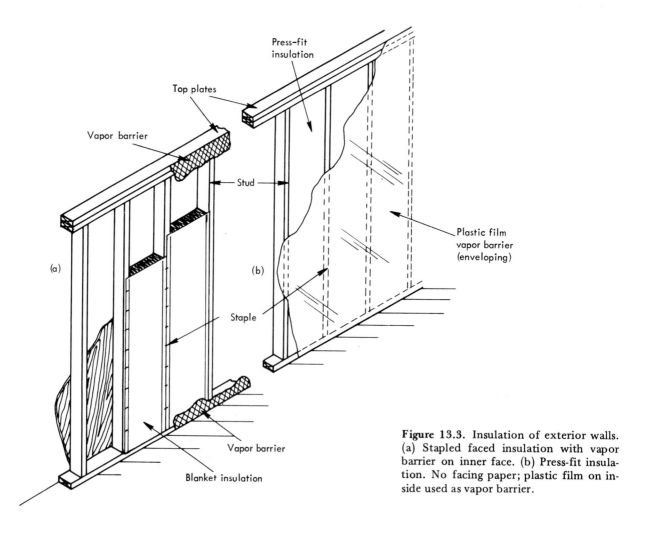

Figure 13.3. Insulation of exterior walls. (a) Stapled faced insulation with vapor barrier on inner face. (b) Press-fit insulation. No facing paper; plastic film on inside used as vapor barrier.

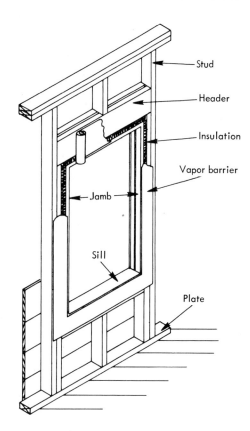

Figure 13.4.
Insulation at a window frame.

front of the reflective material. The space between the strips can be filled with rigid board or a thin blanket.

The opening between door and window frames and the rough framing should also be filled with insulating material. Care should be taken to cover this material with a vapor barrier to prevent possible rotting of the frame by condensed moisture. Figure 13.4 shows how this is accomplished.

13.4.3   Floors and Roofs

Unfinished attic floors can be insulated with loose fill material or with blankets or batts. If blankets or other faced material is used, and if the attic is unheated, a vapor barrier is required on the bottom side of the blanket. Loose fill should be spread evenly and not crushed. The builder should be careful to leave about 3 in. of open space between the perimeter of the attic floor and the outside, for ventilation.

In finished attic or second-floor space, the insulation between roof rafters should be of faced blanket material that is securely stapled to the rafters. The wallboard used for ceilings under the rafters could very well have a reflective surface on one side to act both as extra insulation and as a vapor barrier. The knee walls should also be insulated with blanket and a vapor barrier.

Flat or slightly sloped roofs can be insulated with rigid polyurethane or glass-fiber rigid board laid in pitch or mastic which is then covered with waterproof roofing material.

13.4.4   Heating and Cooling Pipes and Ducts

All lines that carry heat or cold and which pass through exposed spaces where such heat or cold is not required should be insulated. Insulation for piping is usually of air cell pipe covering, which comes in sizes to fit various sizes of piping and is clamped in place with special clamps. Ducts that carry heated or cooled air should be insulated with blanket insulation. There is a question as to whether a vapor barrier is required in such insulation. It will depend on the difference in temperatures and which is normally the warmer side of the insulation.

13.5   PREVENTION OF AIR LEAKAGE

Air leakage through spaces where there are joints between different kinds of materials or through openings such as windows and doors is one of the greatest sources of heat loss. This applies, of course, to air conditioning as well as heating.

These leaks can be stopped by weatherstripping all openings and by caulking. It is interesting to note that a weatherstripped double-hung window will cut down air leakage by one-third and a weatherstripped door will reduce it by one-half. The best kind of weatherstripping to use is the interlocking channel, which can be installed by any skilled carpenter during construction.

Caulking is used when two different materials or pieces of construction meet. Places that need caulking are at joints between window and door drip caps and sills and siding (or masonry, etc.) at joints between window and door frame and siding, at sills where wood structure meets the foundation, at joints where chimney or other masonry meets wood siding, at any break in the exterior wall. All of this caulking can be done with a caulking gun or caulking rope. The work is not very expensive and keeps the structure watertight as well as airtight. Another source of air leakage is the fireplace damper. A good tight damper costs little more than an inefficient one.

Windows are a source of air leakage and have a very low R factor (or a high U factor). A single pane of window glass will, on the average, transmit more than 10 times as much heat as will an ordinary wood or masonry wall. Storm windows will cut this in half, and although it is not usually included in the builder's work, he can certainly recommend the installation as an extra. In general, the agencies that deal in energy conservation have recommended that the glass area of a house be no more than 10 times the floor area. The builder must use judgment while keeping as close to this figure as he can.

13.6   HOW TO DETERMINE MAXIMUM
COST/SAVINGS RATIO

The cost effectiveness of insulation can best be illustrated by a simple example. In Section 13.2 there is an example of an uninsulated house whose exterior walls have a U value of 0.26 or an R value of 4.

In the example such a house will lose 12.5 Btu per hour for every square foot of exterior wall surface. A house of one story that is 50 ft by 25 ft has a perimeter wall of 150 ft. If the exposed wall is 8 ft high, there is an exposed area of 150 x 8 = 1200 ft². If every square foot looses 12.5 Btu per hour, the entire surface will lose 1200 × 12.5 = 15,000 Btu per hour. If the temperature differential of 48°F remains the same for 24 hours, the total loss of heat will be 15,000 × 24 = 360,000 Btu per day. If the house is heated by No. 6 fuel oil each gallon of which contains 140,000 Btu and which costs \$0.50 per gallon, the heat loss will be 360,000 ÷ 140,000 = 2.5 gal per day. 2.5 × 0.50 = \$1.25 per day or \$37.50 per month *through the walls only*. The uninsulated roof will lose much more heat than the walls. The roof area is greater than the wall area, but the same figure is used for convenience. Using a *U* value of 0.29, the loss through the roof for the same period as shown above would be \$37.50 × 0.29/0.26 = \$41.80 per month or a total of \$37.50 + \$41.80 = *\$79.30 per month lost*.* The cost of natural gas and electrical heating is rapidly approaching if not exceeding this figure. The same calculation can be made by using the Btu given by a cubic foot of gas (1100) or by a therm (100,000), depending on how the consumer is billed.

The use of insulation will not reduce this loss to zero but it will reduce it dramatically. For instance, if the walls and rafters are insulated with a rock wool blanket of *R*11 value (see Table 13.1) the following calculation is made:

$$R11 \text{ is equal to a } U \text{ factor of } 1/11 = 0.09$$

This means that the wall loss of heat for the same period as above will be \$37.50 per month × 0.09/0.26 = 0.35 or \$13.12 per month. The heat loss through the roof will be \$41.80 per month × 0.35, or \$14.63 per month.

The saving per month will be \$79.30 – (13.12 + 14.63), or \$51.55 per month. It must be understood that the figures above are averages and that they vary by the severity of the climate, the inside temperature that is maintained, the price of fuel, and other variables, but the calculations as shown need only have different numbers dropped into them to show the value of insulation. Figure 13.5 shows the normal low point of winter temperatures in this country.

As to cost effectiveness, an average figure for the cost of labor and material to install 4-in. fiberglass batts is \$0.47 per square foot. If the total area of walls and roof is taken at 2400 ft², the total cost of labor and material would be 2400 × 0.47 = \$1128. The fuel saving at \$0.50 per gallon and not in a severely cold climate (20°F outside) could be \$51.55 per month for 3 months or \$153.65 plus \$25.00 for 4 months, or \$100.00, for a total of \$254 per year at *present prices* and, of course, with no rationing or shortages. The cost will be saved in a little over 4 years. The author's estimate is that 3 to 3½ years is a more realistic figure.

The foregoing figures have not taken account of air leakage, pipe and duct insulation, and perimeter or underfloor insulation. The cost effectiveness of these is in the same order as stated previously. The owner or builder is strongly advised to install all the insulation. The owner will save the cost in a short period of time, and the builder can increase the price of the house (in speculative building) and probably sell it more readily.

---

*During the year after this was written the cost of fuel oil doubled.

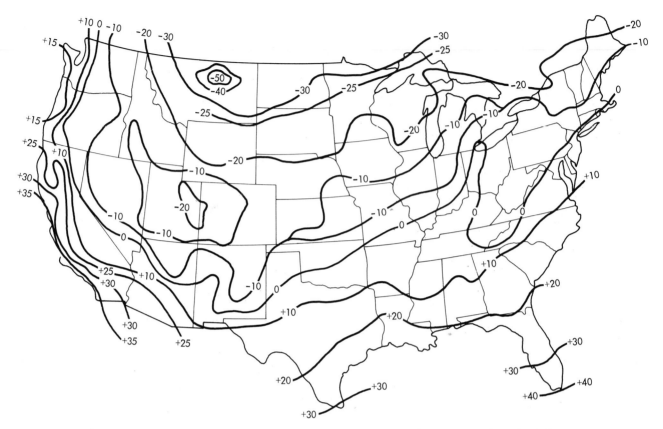

Figure 13.5. The temperatures shown are used as a basis for the design of heating plants and the R value of insulation. (Courtesy U.S. Department of Commerce)

13.7 BUILDING CODES

In the past several years a number of states and municipalities have adopted energy conservation standards. These are very often adopted from the national codes, such as the BOCA or the Uniform Codes.

They set standards for allowable heat losses through walls and roofs and for the structure as a whole. They are being enforced in many localities and many other localities are at the point of adopting and enforcing these standards. The owner, architect, builder, and engineer are advised to inquire about this from the local authority before bids are received or construction is commenced.

CHAPTER FOURTEEN

millwork:
doors, windows, and trim

"Millwork," which is the general topic of this chapter, is the construction name for all finished woodwork that is used in building construction. It includes exterior and interior moldings, exterior and interior doors, windows, door and window frames, mantels, stairways, paneling, shutters, and so on. It does not include framing materials and does not usually include siding or flooring.

14.1 DOORS, EXTERIOR

The choice of the doors for a house is guided by the architecture: Colonial, ranch, contemporary, or other. There are doors to fit almost any normal design.

14.1.1 Choice of Exterior Doors

Front and Service Doors

Main entrance doors should be 1¾ in. thick and made of a kiln-dried close-grained wood such as ponderosa pine and should be treated with a wood preservative. In the panel door which is used on a traditionally designed house, the stiles (the solid vertical members) and rails (the solid horizontal members) should be glued and dowelled and the panels should be 1⅜ in. thick. Figure 14.1 shows a choice of readily available exterior doors. The size of the main door is normally 3 ft wide by 6 ft 8 in. high. Service doors should also be 1¾ in. thick and made of preservative-treated ponderosa pine but can be slightly narrower. Figure 14.2 shows illustrations of readily available service entrance doors which are 2 ft 6 in. wide by 6 ft 8 in. high. Most come with glazed upper panels.

Figure 14.1.
Styles of exterior
front entrance doors.

Figure 14.2. Styles of exterior service doors.

Many contemporary houses are designed for flush doors, which are finished with birch or another decorative veneer. These doors are also 1¾ in. thick and are available as hollow-core or solid-core. Solid-core doors, although more expensive, are strongly recommended to avoid warping, which is caused by differences in moisture content between the interior and exterior. Figure 14.3 shows the construction of flush doors.

It is strongly recommended that all exterior doors not only be weather-stripped but that all exterior entrances in the colder climates be equipped with storm doors (which can come as combination storm and screen). Figure 14.4 shows a cross section of a typical door casing and door as well as a storm and screen door.

Figure 14.3. Flush doors:
(a) solid-core door;
(b) hollow-core door.

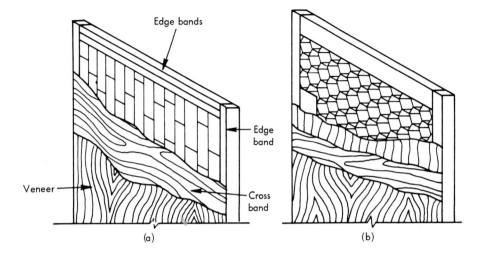

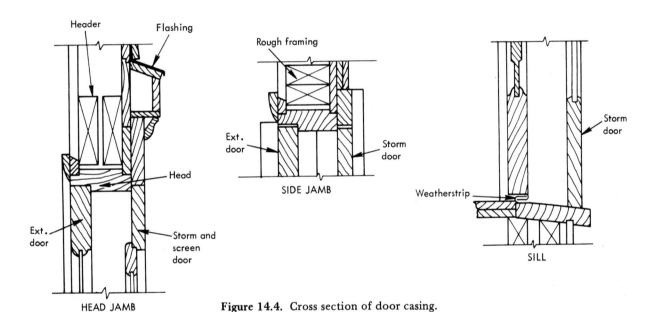

Figure 14.4. Cross section of door casing.

Sliding Doors

Another type of exterior door which has come into increasing use is the sliding door. This type of door is used to a great extent as a replacement for the so-called French doors, whose purpose was to provide a wide opening and glass area between indoor and outdoor living. Sliding doors are usually of aluminum, but there are excellent doors made of wood, as shown in Figure 14.5. Because of the large glass areas, all sliding doors should be of $\frac{1}{4}$ in. plate, of float glass, or (for energy saving) be double-glazed. Sliding screens are also available for all such doors. The builder is advised to size the lintels over the wide openings required so that there will be no sag, which can seriously affect the usefulness of the doors. Steel-angle lintels are suggested for masonry construction.

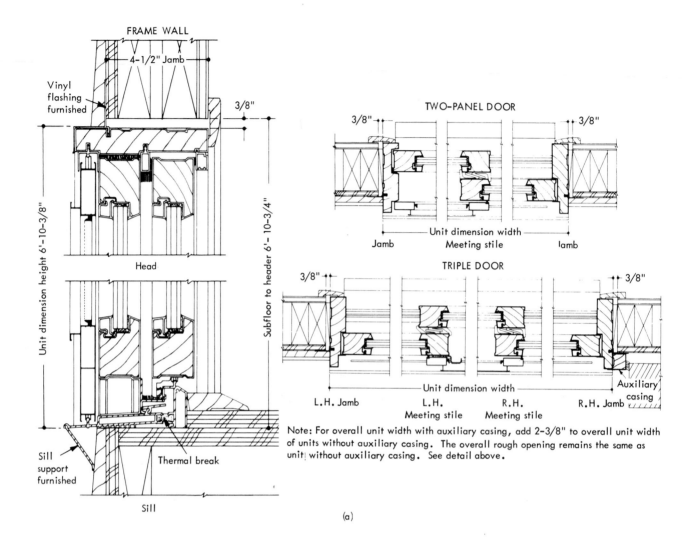

FRAME WALL

4-1/2" Jamb

Vinyl flashing furnished

3/8"

Unit dimension height 6'-10-3/8"

Subfloor to header 6'-10-3/4"

Head

Sill support furnished

Thermal break

Sill

TWO-PANEL DOOR

3/8" 3/8"

Unit dimension width

Jamb   Meeting stile   Jamb

TRIPLE DOOR

3/8" 3/8"

Unit dimension width

Auxiliary casing

L.H. Jamb   L.H.   R.H.   R.H. Jamb
          Meeting stile Meeting stile

Note: For overall unit width with auxiliary casing, add 2-3/8" to overall unit width of units without auxiliary casing. The overall rough opening remains the same as unit without auxiliary casing. See detail above.

(a)

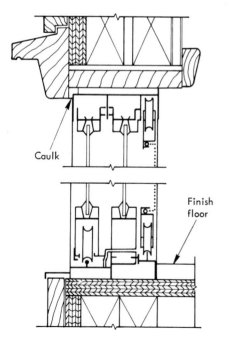

Caulk

Finish floor

Figure 14.5. Sliding doors: (a) wood, vertical and horizontal cross sections; (b) aluminum, vertical and horizontal cross sections. (Courtesy Andersen Corporation, Bayport, Minn. 55003.)

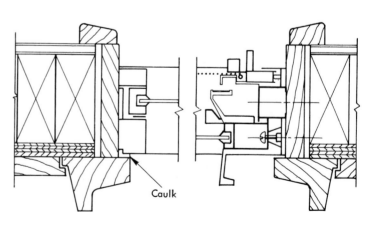

Caulk

(b)

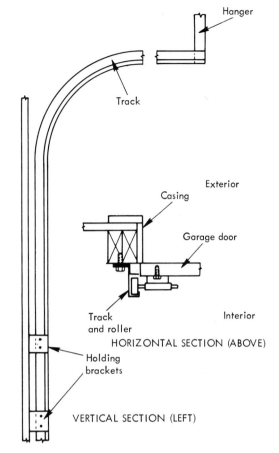

Hanger

Track

Exterior

Casing

Garage door

Interior

Track
and roller

HORIZONTAL SECTION (ABOVE)

Figure 14.6. Tracks for
sectional overhead door.

Holding
brackets

VERTICAL SECTION (LEFT)

14.1.2   Garage Doors

A widely used garage door is the wood sectional overhead door, which is
mounted on rollers riding in tracks on both sides and which is counter-
weighted by powerful springs so that it can be manually raised and lowered,
as shown in section in Figure 14.6. There are also single-section swinging
garage doors as well as single-section doors actuated by side springs. The
sectional roll-up doors come in a number of sizes and patterns, as shown
in Figure 14.7. This type of door can also be made to be moved by an
electric motor, which can be actuated by a button or by remote radio
control so that it can be opened by the car driver as he or she approaches
the garage.

   The builder is cautioned about the size of the lintel over the wide opening of a garage door. He must remember that this lintel is (in most cases)
supporting the roof and a possible snow load. Local codes will give the
proper sizes for this supporting beam.

Figure 14.7. Patterns of sectional overhead doors: 8, 9, 10, 17, and 18 ft
wide, 6'-6" to 7'-0" high (17-ft and
18-ft doors not shown).

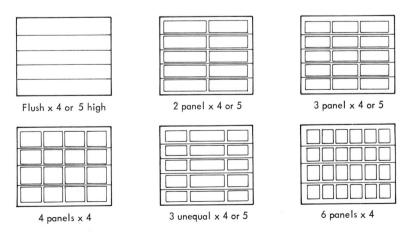

Flush x 4 or 5 high

2 panel x 4 or 5

3 panel x 4 or 5

4 panels x 4

3 unequal x 4 or 5

6 panels x 4

14.2 WINDOWS

In the great majority of cases, windows are factory-made and factory-assembled and come to a construction job as a complete unit with frame and sash, ready to be installed in the rough openings that have been prepared for them. As in the case of exterior doors, the size and style of the windows should be chosen with a view to the appearance and style of the house, and also of course, by the owner's preference. In the case of a speculative builder, it is well to choose a type and size of window that is standard for the type of architecture and for the area.

Windows serve as a means of natural light and ventilation in residential construction. As this has an effect on the safety and welfare of the occupants, it is mentioned in all building codes. A typical code requirement is as follows: "Window size: Windows and exterior doors may be used as a natural means of light and ventilation, and when so used their aggregate glass area shall amount to not less than one-tenth of the floor area served, and with not less than one-half of this required area available for unobstructed ventilation."

14.2.1 Types and Sizes of Windows

Windows are available in many types, all of which are used extensively to satisfy owner's preference, to provide light and air as required, to efficiently shelter the interior from the elements, and to be, as far as possible, maintenance-free. Wood windows are made of decay-resistant clear grade or heartwood of ponderosa pine, certain grades of spruce, redwood, cypress, and cedar. In the temperate climate that prevails over most of this country, the great preponderance are of pine or spruce—the other woods being much more expensive.

Double-Hung Windows

This type of window has for a long time been one of the most widely used in residential construction. It consists of a frame into which are fitted an upper and lower sash which can slide by each other and then give a maximum opening of one-half the entire opening. The sash is counterbalanced by springs or counterbalances so that it will not be too difficult to slide and so that it will stay in place without movement. The sash is made of $1\frac{3}{8}$-in.-thick stock at the sides, the top and the bottom, and at the meeting rails, which are at the top of the bottom sash and the bottom of the top sash. The meeting rails are so designed that they will fit together to form a tight joint. Figure 14.8 shows cross sections, both horizontal and vertical, of a typical wood double-hung window design. The readily available sizes of double-hung windows which are in stock at most material dealers is as follows: 2'-0", 2'-4", 2'-8", and 3'-0" wide, all in heights of 3'-2", 3'-10", 4'-2", 4'-6", and 5'-2".

These windows can be obtained as 1 over 1, two single-glazed sash; 2 over 2, which is often used in contemporary houses, each sash having one horizontal muntin (which is the thin wood strip which divides the glazing); 6 over 6, which is a typical window used in traditional houses and which

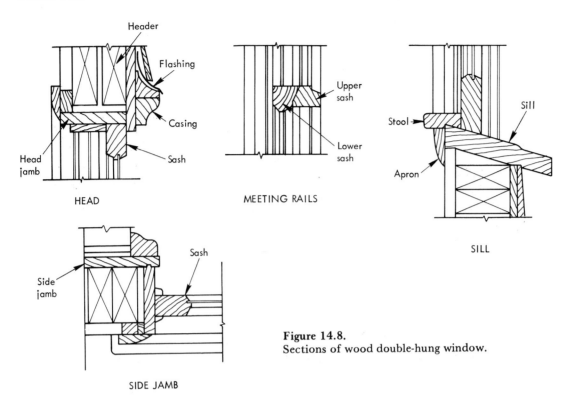

Figure 14.8.
Sections of wood double-hung window.

has six panes of glass divided by muntins in each sash; or 6 over 1, which contains an upper sash with six panes and a lower sash with a single pane.

There are many other sizes available, as can be seen from Figure 14.9. The windows shown can be adapted to any style of architecture by the use of snap-in muntins, as shown.

Double-hung windows are also available in aluminum. The essential features are the same as for wood. Although aluminum is maintenance-free as far as painting is concerned, it should be treated to avoid surface discoloration and there is a certain amount of corrosion due to air pollution and other causes. Also, the heat loss through a metal window is much greater than through wood.

Horizontal Sliding Windows

As the name implies horizontal sliding windows are made to slide by each other just as are double-hung windows. They come fully factory assembled and in a variety of sizes, as shown in Figure 14.10. The windows shown here are permanently covered with rigid vinyl, which makes them maintenance-free, but there are sliding windows available which can be painted to suit an owner's or builder's preference. As for all wood windows, sliding windows should be treated with preservative. Sliding windows are usually double-glazed and the frames can be fitted with stationary screens or storm sash.

As for double-hung windows, the sliding window opens to a full one-half of the window opening. These windows are used in better-grade contemporary houses and of course can be used anywhere. They are generally more expensive than are double-hung windows.

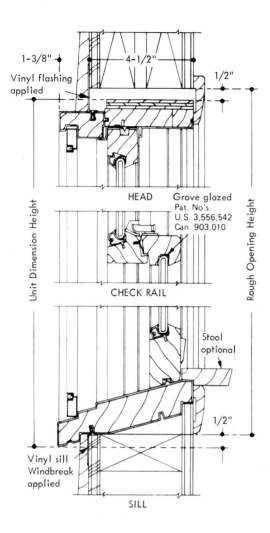

HEAD

1-3/8"　4-1/2"　1/2"

Vinyl flashing applied

Unit Dimension Height

Rough Opening Height

Grove glazed Pat. No's.
U.S. 3,556,542
Can. 903,010

CHECK RAIL

Stool optional

1/2"

Vinyl sill Windbreak applied

SILL

Figure 14.9. Section and sizes of double-hung windows. (Courtesy Andersen Corporation, Bayport, Minn. 55003)

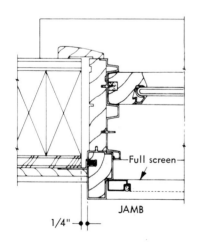

Full screen

JAMB

1/4"

BASIC UNITS

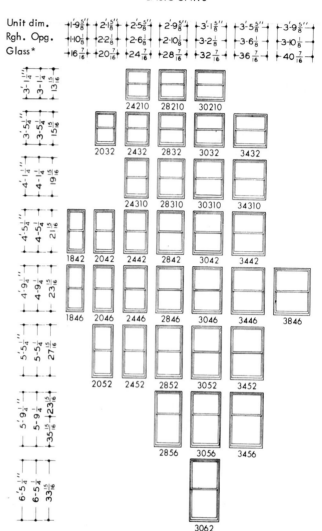

*Unobstructed Glass Sizes shown in inches.
Glass height for one sash only.

PATTERNS OF SNAP-IN GRILLES AND MUNTINS

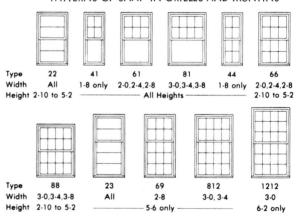

Type	22	41	61	81	44	66
Width	All	1-8 only	2-0,2-4,2-8	3-0,3-4,3-8	1-8 only	2-0,2-4,2-8
Height	2-10 to 5-2		— All Heights —			2-10 to 5-2

Type	88	23	69	812	1212
Width	3-0,3-4,3-8	All	2-8	3-0, 3-4	3-0
Height	2-10 to 5-2		— 5-6 only —		6-2 only

VERTICAL DETAIL

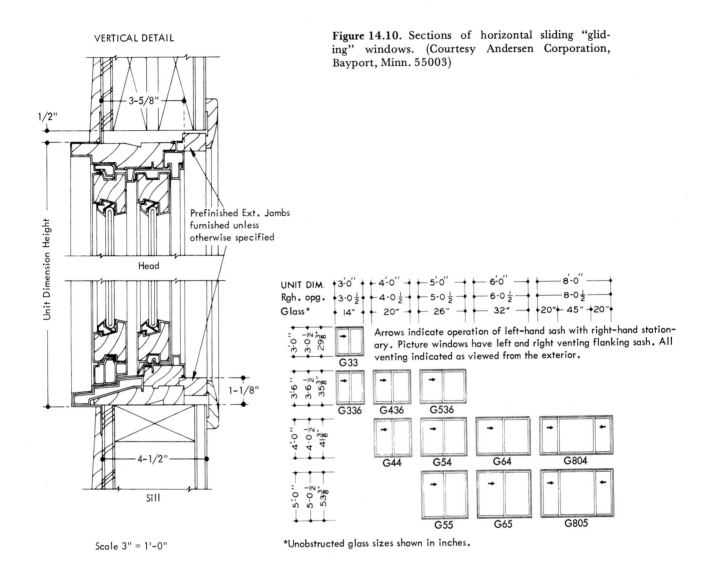

3-5/8"

1/2"

Unit Dimension Height

Head

Prefinished Ext. Jambs
furnished unless
otherwise specified

1-1/8"

4-1/2"

Sill

Scale 3" = 1'-0"

Figure 14.10. Sections of horizontal sliding "gliding" windows. (Courtesy Andersen Corporation, Bayport, Minn. 55003)

UNIT DIM.   3'-0"   4'-0"   5'-0"   6'-0"   8'-0"
Rgh. opg.   3-0½   4-0½   5-0½   6-0½   8-0½
Glass*   14"   20"   26"   32"   20" 45" 20"

Arrows indicate operation of left-hand sash with right-hand stationary. Picture windows have left and right venting flanking sash. All venting indicated as viewed from the exterior.

G33

G336   G436   G536

G44   G54   G64   G804

G55   G65   G805

*Unobstructed glass sizes shown in inches.

TABLE OF AVAILABLE SIZES

HORIZONTAL DETAIL

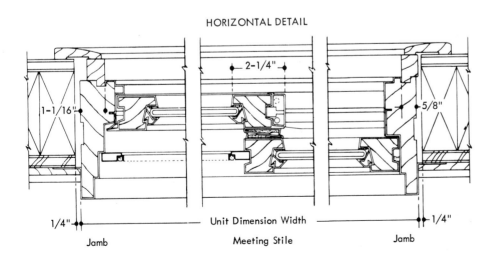

2-1/4"

1-1/16"

5/8"

1/4"   Unit Dimension Width   1/4"

Jamb   Meeting Stile   Jamb

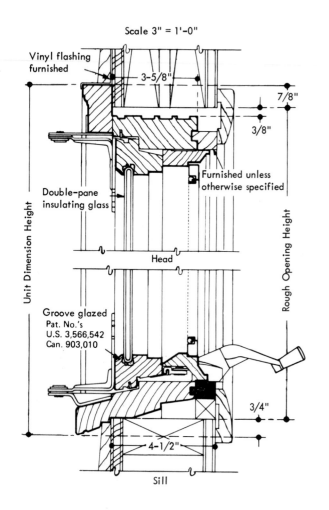

Scale 3" = 1'-0"

Vinyl flashing furnished

3-5/8"

7/8"

3/8"

Furnished unless otherwise specified

Double-pane insulating glass

Head

Unit Dimension Height

Rough Opening Height

Groove glazed
Pat. No.'s
U.S. 3,566,542
Can. 903,010

3/4"

4-1/2"

Sill

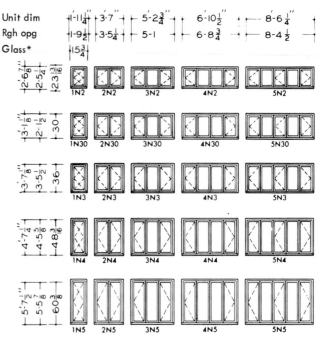

*Unobstructed glass size in inches.

TABLE OF AVAILABLE SIZES

Figure 14.11. Wood casement windows. (Courtesy Andersen Corporation, Bayport, Minn. 55003)

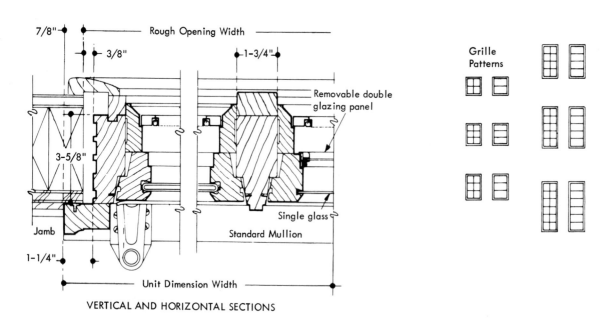

7/8"

Rough Opening Width

3/8"

1-3/4"

Removable double glazing panel

3-5/8"

Single glass

Jamb

Standard Mullion

1-1/4"

Unit Dimension Width

VERTICAL AND HORIZONTAL SECTIONS

Grille Patterns

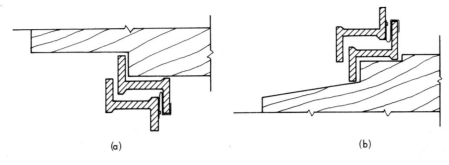

(a)   (b)

(c)

Figure 14.12. Sections of a metal casement window; (a) head; (b) sill; (c) side jamb.

Casement Windows

Casement windows have side-hinged sash which are usually designed to swing outward. The sash is opened and closed by a crank handle on the inside which actuates a swing arm attached to the sash. Casement sash is very effectively used in certain types of architecture and can be very useful in certain places in a residence, such as over a kitchen sink, when a simple turn of a crank will open or close the window. Casement windows are widely used in residential construction in Europe. The windows always come ready assembled from the manufacturer. The wood casements are pre-treated with wood preservative and the metal ones (aluminum and steel) are made weather-resistant by anodizing for aluminum and by preservative weatherproof coatings for steel. All casements are pre-weather-stripped and can be fitted with inside screens or storm sash. Figure 14.11 shows sections of a wood casement window and gives a choice of sizes. One advantage of casement windows is that they can be opened to the full extent of the window opening.

Steel casements are standardized by The Steel Window Institute, 2130 Keith Building, Cleveland, Ohio 44115. Some standard sizes for single casements are as follows: $1'-8\frac{7}{8}''$ by $2'-9''$, $4'-1''$, $5'-5''$, and $6'-9''$ high and for double casements $3'-4\frac{7}{8}''$ or $4'-0\frac{7}{8}''$ wide. Figure 14.12 shows a cross section of a metal window.

Aluminum windows are standardized by the Architectural Aluminum Manufacturers Association, 35 East Wacker Drive, Chicago, Ill. 60601.

Stationary Windows

Stationary windows are usually used in combination with movable windows, such as a picture window flanked on either side by a double-hung or casement window. The stationary window consists of a frame and a sash which holds either a single large light of glass or which is divided by muntins into a number of smaller lights. The sash is fixed permanently to the frame and cannot be moved. Because of the usual large expanse of glass, the wood members of a stationary window are at least $1\frac{3}{4}$ in. thick to provide sufficient strength. Stationary windows very often form the center piece of a bay window. Also, because of the large expanse of glass, it is recommended that a stationary window be double-glazed. In some instances a stationary window can be used without a sash. The glass can be set directly into rabbeted members of the house frame and secured with stops and back and face puttying.

Awning Windows

An awning window is one in which the movable units are hinged horizontally at the top and which swing outward. There can be a single movable sash or a series of one above the other separated by horizontal transom bars. Such a series of sash can be made to move in unison. The jambs and transom bars must be strong enough to support the weight of the outward swinging sash. In some cases it serves a better purpose if the sash is hinged at the bottom and swings in. This is called a hopper window. Both types protect the interior from weather while they are open.

Awning windows or hopper windows are available in wood, steel, and aluminum. Figure 14.13 shows the details of a wood awning window that is operated by a crank handle. Both awning and hopper windows are available to be manually operated.

Figure 14.13. Details of wood awning windows: (a) head; (b) meeting rail; (c) sill; (d) awning narrow mullion; (e) awning support mullion; (f) awning transom. (Courtesy Andersen Corporation, Bayport, Minn. 55003)

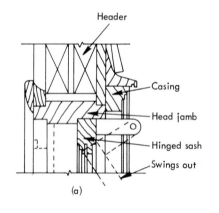

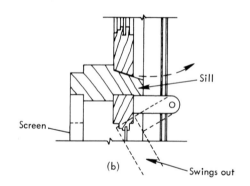

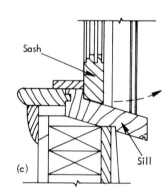

Joining basic units to form multiple units or picture window combinations without support between units.

Joining basic units to form multiple units or picture window combinations with a 2 x 4 vertical support between units.

Joining basic units by stacking units to form combinations.

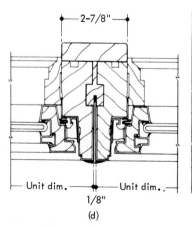

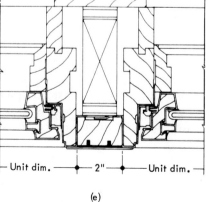

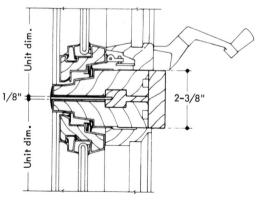

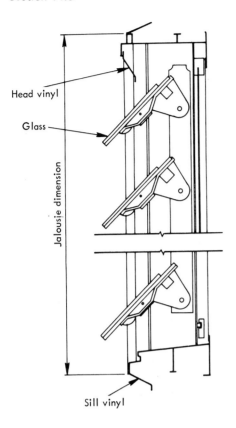

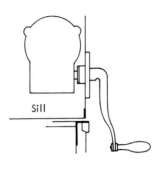

Figure 14.14.
Section of jalousie window.

Jalousie Windows

Jalousie windows consist of a series of horizontally hinged overlapping glass louvers which pivot in unison. This type of window is used for porches or closed-in patios. It keeps out rain but it is not otherwise weathertight and it is inadvisable to use it as a window for any space that must be heated. Jalousie windows are more often used in warm climates for maximum ventilation in non-air-conditioned spaces. Figure 14.14 shows a cross section of a jalousie window.

14.3　TRIM, EXTERIOR

Trim, which is one of the subdivisions of millwork, is the lighter woodwork in the finish of a building or, specifically, the finish materials in a building, such as moldings applied around doors and windows on the exterior or interior and at floors and ceilings on the interior. Contemporary houses do not use as much trim as traditional houses with their cornices and moldings and ornamental railings. Materials used for trim should be of the better grades of softwoods with clear grain and easy workability. In places where the trim is subject to continuous moisture decay resisting woods, such as redwood, cypress, and cedar are recommended. A moisture content of not more than 12% is recommended in most areas except in the dry Southwest, where it should be 9%.

Most trim is cut and fitted on the job, but such items as railings, shutters, columns, and so on, are usually factory-assembled ready for installation. It is recommended that all preassembled trim be predipped in a wood preservative before being sent to the job. Most manufacturers do this and the builder should make sure that it has been done. Exterior trim should be fastened with rust-resistant nails or screws and even when fitted on the job it is a wise precaution to end-dip the members before fitting them together. A standard siding nail or a finish nail is used in most cases and the nail should be set below the surface and puttied over before painting to avoid staining of the surface.

The most important exterior trim occurs where the roof of a structure meets the exterior walls. This is an important junction point both from the construction and architectural viewpoints. A well-built cornice at both the eaves and the gable ends is important not only as a protection against weather such as ice, snow, or rain but also can add to the appearance and saleability of a house.

The meeting of the roof and the walls at the eaves can end in a narrow or wide box cornice formed by overhanging rafters which are boxed in by a soffit board and a fascia board, or it can end as a simple close cornice in which the roof shingles overhang a frieze board and a simple molding, as shown in Figure 14.15.

The meeting of the roof and the walls at the gable ends is usually finished with a fascia board that is overhung by the roof covering. In more elaborate finishes there can be an overhanging cornice, as shown in Figure 14.16.

14.4   DOORS AND TRIM, INTERIOR

It is strongly recommended that interior doors and trim not be installed until a house is weathertight and dry with all exterior windows, doors, and trim in place; the finish floor laid; and the heat on in cold or wet weather. The moisture content of interior trim should not be over 11% and as low as 6% in the dryer sections of the country.

14.4.1   Doors

Interior doors come in various patterns and are manufactured of a variety of woods. There are hollow-core flush doors, one-panel doors, cross-panel doors, six-panel Colonial doors, and louvre doors. Interior doors can be hung to slide, to fold open, or to swing on hinges. They can be made of ponderosa pine, white pine, or spruce. Better doors are made of redwood, birch, or gum. When a natural finish is required, there are doors of walnut, birch, cherry, or other woods with desirable colors and grains. Figure 14.17 shows some interior door patterns.

The usual interior swing door is $1\frac{3}{8}$ in. thick and the stiles (the vertical members) and rails (the horizontal members) should be of this thickness. The panels are set into these solid members, which should be fastened by notching and glueing. Louvre doors are usually $1\frac{1}{8}$ in. thick and are used mostly for closets.

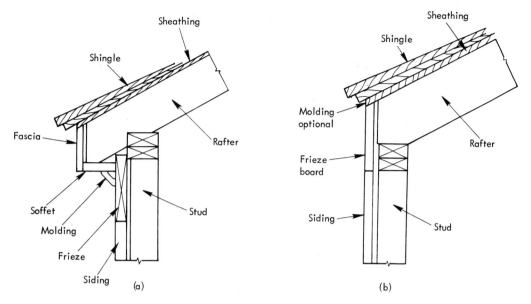

Figure 14.15. Examples of eave cornices: (a) box; (b) close.

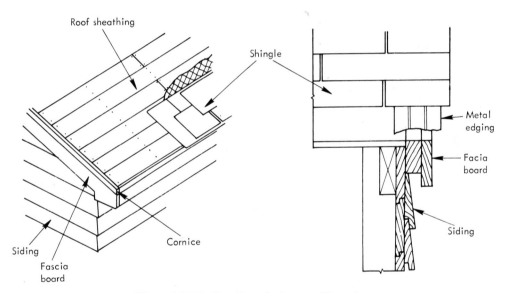

Figure 14.16. Samples of trim at gable ends.

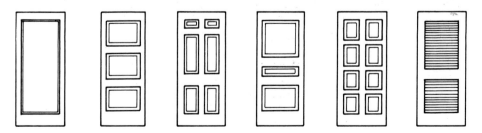

Figure 14.17. Patterns of interior doors.

235

Hardware for swing doors consists of hinges, lock or latch sets, strike plates, and door stops. This finish hardware comes in many finishes, such as brass, bronze, and nickel. Many lock or latch sets to closets or bathrooms have glass knobs or ornamental metal knobs.

The heavy exterior doors are hung on three hinges each 4 by 4 in., and the lighter interior doors are hung on two hinges which can be 3½ by 3½ in. Loose pin butt hinges should be used so that the door can be removed by removing the pin. Lock sets for exterior doors come in many prices and in many patterns. A good cylinder lock with a latch and a dead bolt is strongly recommended. Interior doors can be fitted with either latch sets, which is usual, or lock sets, where rooms have to be kept secure.

The strike plate is the metal plate which is secured to the door frame and which receives the latch or dead bolt and therefore secures the door. The jamb of the door frame is routed out to receive the strike plate. The builder must be very careful that the dead bolt or latch fit exactly into the strike plate.

The rough door frame is made when the interior partitions are framed. The rough opening is made 2½ in. wider and 3 in. higher than the finish door. This allows for the finish door frame, which is made of two side jambs, a head jamb, and a stop molding. The jamb pieces come in standard widths to fit over plaster or wallboard interior finishes. They can be fitted on the job or they can come assembled with the door already hung. The head jamb fits into a notch in the side jambs and the joint is nailed and glued. The carpenter uses wood wedges between the rough opening and the finish jamb to obtain a true fit for the door. Figure 14.18 shows the fitting of a head jamb to a side jamb and also shows the finishing of the door frame with the stop and the casing.

Figure 14.18. Construction of door frame.

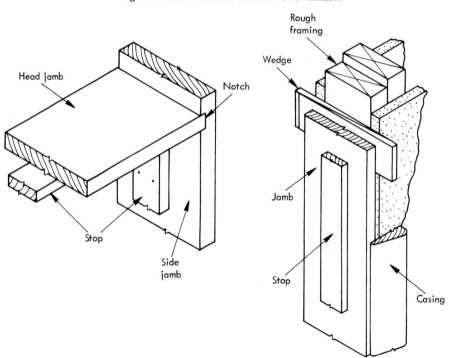

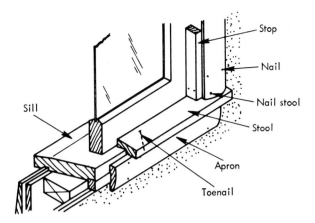

Figure 14.19.
Interior window trim.

14.4.2   Window Trim

Windows come completely assembled with sash and frames ready to be fitted into the rough openings left for them in the framing. The only trim necessary for the ordinary window opening is the exterior casing, which serves as a stop for the exterior siding, and the interior casing, which serves as a finish between the window frame and the interior wall. The interior of a window opening is also finished with a sill or stool and apron, as shown in Figure 14.19. When window trim is installed, the stool is nailed in place first and it extends beyond the window opening so that it serves as a stop for the side casing and the window stop. As can be seen in the illustration, the stool fits against the window to serve as a weather stop. The casing and apron are then installed against the stool.

14.4.3   Ceiling and Base Trim

The only other trim ordinarily used in interiors besides the door and window trim is a ceiling or cornice molding and base molding (Figure 14.20). Both these items come in various sizes and shapes. Many houses do not have a ceiling molding and only a very simple base molding, but there are stock moldings available for more elaborate designs. Many better homes have a chair rail in the living room or dining room to serve as a divider between the upper and lower part of the wall. The chair rail is a decorative molding which is often used to separate paint from wallpaper or two colors of paint or can simply be used as a decoration.

Figure 14.21 shows a variety of moldings and their sizes.

Figure 14.20. Base molding: (a) outside corner; (b) inside corner.

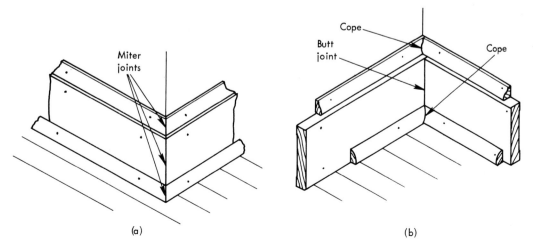

(a)

(b)

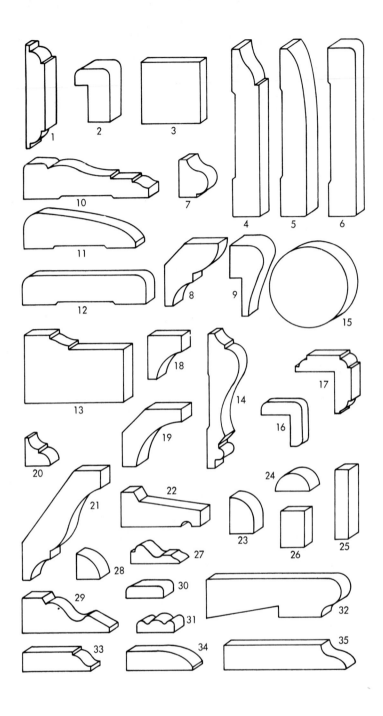

Figure 14.21.
Stock interior and exterior moldings.

No.	Kind	Size
1	Astragal	3/8 x 1-3/8
	"	3/8 x 1-5/8
	"	3/8 x 2
2	Back band	11/16 x 1-1/16
3	Balluster	3/4 x 3/4
	"	1-1/8 x 1-1/8
	"	1-5/16 x 1-5/16
	"	1-5/8 x 1-5/8
4	Base colonial	9/16 x 3-1/2
5	Base clamshell	9/16 x 3-1/2
6	Base sanitary	11/16 x 3-1/2
7	Base mld.	11/16 x 11/16
8	Bed mld.	11/16 x 1-5/8
	"	11/16 x 2-1/4
	"	11/16 x 2-5/8
9	Cap	11/16 x 1-1/8
10	Casing, colonial	11/16 x 2-1/2
11	Casing, clamshell	11/16 x 2-1/4
12	Casing, sanitary	11/16 x 2-1/2
13	Casing, exterior	1-5/16 x 2
14	Chair rail	11/16 x 2-1/2
15	Closet pole	1-1/8
	"	1-3/8
	"	1-5/8
16	Corner guard	7/8 x 7/8
17	Corner guard	1-1/8 x 1-1/8
	"	1-3/8 x 1-3/8
18	Cove mld.	5/8 x 3/4
	"	11/16 x 7/8
19	Cove mld.	9/16 x 1-5/8
	"	11/16 x 2-5/8
20	Cove & bead	3/8 x 1/2
	"	5/8 x 3/4
21	Crown	11/16 x 2-5/8
	"	11/16 x 3-5/8
	"	11/16 x 4-5/8
22	Drip cap	1-1/16 x 1-5/8
23	Floor mld.	1/2 x 3/4
24	Half round	1/4 x 1/2
	"	5/16 x 5/8
	"	3/8 x 3/4
	"	3/8 x 7/8
25	Lattice	1/4 x 1-1/8
	"	1/4 x 1-3/8
	"	1/4 x 1-5/8
	"	1/4 x 2
26	Parting strip	1/2 x 3/4
27	Panel mld.	3/8 x 1
28	Quarter round	1/4 x 1/4
	"	3/8 x 3/8
	"	1/2 x 1/2
	"	5/8 x 5/8
	"	3/4 x 3/4
	"	7/8 x 7/8
29	Rake mld.	11/16 x 1-5/8
	"	11/16 x 2-1/2
30	Screen mld.	1/4 x 3/4
31	"	1/4 x 5/8
32	Stool, window	11/16 x 3-1/4
33	Stop, colonial	7/16 x 7/8
	"	7/16 x 1-1/8
	"	7/16 x 1-3/8
34	Stop, clamshell	7/16 x 7/8
	"	7/16 x 1-3/8
35	Stop, garage door	11/16 x 2-1/2

CHAPTER FIFTEEN

chimneys, fireplaces, and stairs

15.1 CHIMNEYS AND FIREPLACES

Any house that uses fossil fuel for heating must have a means of getting rid of the product of combustion. This means the construction of a chimney. The chimney must be built so that (1) it is insulated from any surrounding combustible material such as wood framing, (2) it does not leak any gases into the houses (such gases include carbon monoxide, which is deadly), and (3) it is large enough to take care of the amount of gases and other products of combustion that may be produced.

15.1.1 Size and Construction of Chimneys

As chimneys can be a safety hazard or a fire hazard if they are improperly built, they are mentioned in building codes and there are minimum requirements for them. In many parts of the country with mild climates the less expensive houses may not have a masonry chimney but simply a flue, which may be made of prefabricated metal. Such flues are allowed by codes provided that they are approved by Underwriters Laboratories or other recognized fire safety organization. The builder must be specially careful to insulate such flues from any surrounding combustible material. These flues can be used for fireplaces as well as for heating furnaces. There are many houses built with fireplaces away from the heating plant so that a metal flue is the only practicable means of carrying away the smoke and fumes. Such a flue must be fitted with a large metal hood that covers the entire area of the fireplace.

In most parts of the country chimneys are made of masonry and usually contain at least two flues—one for the furnace and one for a fireplace. The masonry chimney must be self-supporting and must be supported on a foundation as approved by code. The size of the chimney is determined by the size of the flue lining, which is made of kiln-fired fire clay. Such clay linings are available as round or rectangular shapes. Table 15.1 gives some standard sizes. They all come in 2-ft lengths.

TABLE 15.1. Sizes of Clay Chimney Linings

Round Flue			Rectangular Flue		
Diameter	Area	Thickness	Dimensions	Area	Thickness
6	26	$\frac{5}{8}$	$4\frac{1}{2} \times 8\frac{1}{2}$	22	$\frac{5}{8}$
8	47	$\frac{3}{4}$	$4\frac{1}{2} \times 13$	36	$\frac{5}{8}$
10	74.5	$\frac{7}{8}$	$8\frac{1}{2} \times 8\frac{1}{2}$	51	$\frac{5}{8}$
12	108	1	$8\frac{1}{2} \times 13$	79	$\frac{3}{4}$
15	171	$1\frac{1}{8}$	$8\frac{1}{2} \times 18$	108	$\frac{7}{8}$

[a]All dimensions in inches. Areas in square inches. Round flue diameter is inside clear. Rectangular dimensions are outside.

Standard practice calls for the flue lining to be covered with at least 4 in. of masonry and flue sizes are adapted for the purpose so that an $8\frac{1}{2}$- by 13-in. flue lining will require seven common bricks to surround it. Many builders do not use flue lining all the way to the top of the chimney; this is poor practice, because the top is the part of the chimney that is subject to winds and water and freezing, and such action can cause interior chimney collapse, with possibly serious consequences. When the chimney is being built, the flue lining should go up slightly ahead of the masonry so that it can be thoroughly slushed in with mortar (which should be cement, not lime mortar) and also so that the joints between the flue sections can be

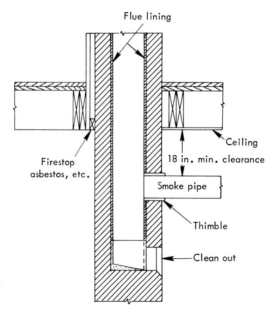

Figure 15.1. Cross section of furnace smoke pipe inserted into chimney.

smooth. The flue lining should start below the "thimble," which is the place where the smoke pipe from the furnace enters the masonry chimney and should continue up and extend above the top of the chimney. The thimble can be a cast-iron ring or simply fire clay, which is carefully packed around the flue pipe. The metal smoke pipe from boiler to chimney must be more than 18 in. below the ceiling above the heating plant and the chimney must extend a certain distance above the ridge of the roof as set by local code. If there is more than one flue, the separate flues should be separated by 4 in. of brick.

Figure 15.1 shows how a furnace flue is inserted into a chimney, Figure 15.2 shows the top of a two-flue chimney, and Figure 15.3 shows wood framing around a chimney. Sometimes a smoke flue must go through a combustible partition. Figure 15.4 shows how this can be accomplished in accordance with code requirements.

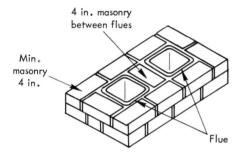

Figure 15.2. Top of two-flue chimney. Note use of whole bricks.

Figure 15.3. Wood framing around a chimney. Note minimum clearances.

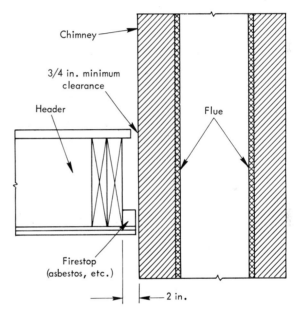

Figure 15.4. Smoke pipe through a combustible partition.

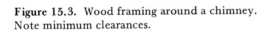

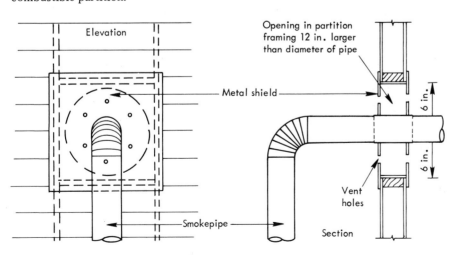

15.1.2 Fireplaces

In the great majority of cases, the fireplace is built more as a decorative feature of the house than as a source of heat. The ordinary source of heat, such as a radiator or a hot-air register, circulates warm air. In the case of the radiator, the hot surface warms air, which rises and then falls as it cools. The hot surface also radiates heat. The hot air register blows warm air into the room, which rises and then falls as it cools. The fireplace does not circulate heat but simply heats by radiating heat from the hot brick surfaces of its sides and rear. As a matter of fact, the fireplace draws air from the room, heats it, and then sends it up the chimney. The owner, architect, or builder can design a fireplace with a built-in steel lining which will heat air that is drawn in through registers at the sides of the fireplace opening and discharge warm air through registers in the chimney breast at a point in the upper part of the wall. In this way the fireplace can be made much more efficient as a heating device.

Design of a Fireplace

Although the fireplace is not a heat-efficient device, it can be made as efficient as possible by proper design. In any case, it must "draw" well or it will become a smoky nuisance. There are certain well-tested formulas for obtaining the proper ratios between the height and depth of the fireplace opening, the size of the flue, the size of the throat, and the width of the back and sides. Table 15.2 gives these ratios and Table 15.3 summarizes them. It is to be noted that the dimensions given allow some leeway, because height of chimney, wind direction, wind velocity, location of the fireplace in the house are all variables that do not allow any positive dimensions.

TABLE 15.2. Fireplace-Opening Height/Depth Ratios

Depth of fireplace	$\frac{2}{3}$ the height of the opening
Area of flue	$\frac{1}{10}$ the area of the fireplace opening[a]
Width of back wall	6 to 8 in. narrower than front
Area of throat	$1\frac{1}{4}$ to $1\frac{1}{3}$ times the area of the flue

[a]For chimney more than 15 ft high. For lower chimneys, this should be $\frac{1}{8}$.

TABLE 15.3. Summary of Fireplace Dimensions

Width, w	Height, h	Depth, d	Minimum Width Back Wall, c	Height of Back Wall, a	Height, Including Back Wall, b	Flue Size External	Flue Size Internal
28	24	16–18	18	14	16	$8\frac{1}{2}$ × 13	10
30	28-30	18-20	20	14	18	$8\frac{1}{2}$ × 13	10
36	28-30	18-20	24	14	18	$8\frac{1}{2}$ × 13	12
42	28-32	18-20	30	14	18	13 × 13	12

Some of the dimensions given in Table 15.3 are minimums, especially for the depth and the width of the back wall. The ratios listed in Table 15.2 are much more strongly recommended. The dimensions in Table 15.3 are to be used in conjunction with Figure 15.5.

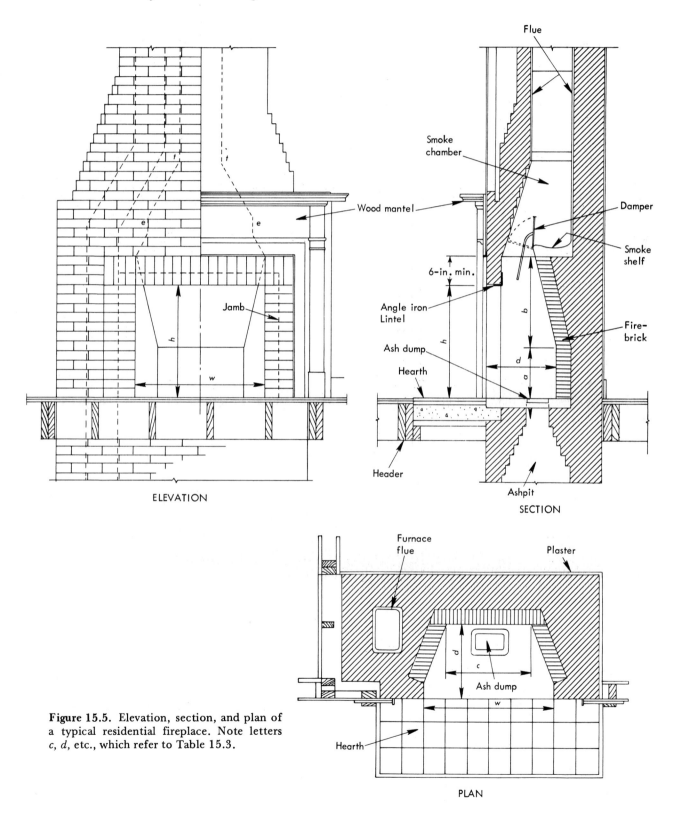

Figure 15.5. Elevation, section, and plan of a typical residential fireplace. Note letters *c, d,* etc., which refer to Table 15.3.

Construction of a Fireplace

As mentioned in Section 15.1, the masonry chimney which contains the fireplace flue must by code be self-supporting and must rest on a foundation that is of sufficient dimensions to properly support it.

The hearth of a fireplace should be made of noncombustible material such as brick, stone, or terra cotta, and it should extend at least 20 in. from the chimney breast and extend at least 12 in. on either side of the fireplace opening. It should be supported (preferably) on fire-resistant construction. The hearth can be flush with the floor or raised, as is now being done in contemporary houses.

The walls of the fireplace must be lined with firebrick backed with 8 in. of masonry (Code). The lining can also be of steel at least ¼ in. thick and this must be backed with 8 in. of masonry.

The fireplace front or the jambs should be wide enough to be pleasing architecturally but must also be so designed that combustible material such as a mantel is kept at a safe distance from the opening. The woodwork of the mantel must be at least 6 in. from the sides and 12 in. from the top. Depending on the size of the fireplace opening, a good proportion should be observed. Many fireplace jambs are decorated with glazed tile or face brick.

The top of the fireplace opening is supported by a steel lintel, usually an angle iron. For an opening less than 48 in. wide, a 3½ - by 3½ - by ¼ -in. angle is sufficient.

The throat is at the top of the fireplace, where the side and rear walls curve inwardly to guide the smoke and fumes up the chimney. It should be 8 in. above the top of the fireplace opening and should be 1¼ to 1⅓ larger in area than the flue. The flue lining starts about 8 in. above the throat.

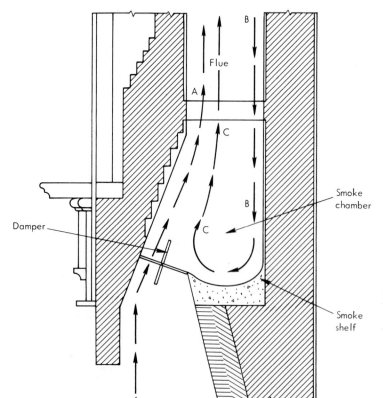

Figure 15.6. A fireplace smoke shelf in action. Smoke from the fireplace goes straight up the flue (A). A gusty wind may blow the smoke back down the chimney (B). The smoke shelf diverts it and sends it back up (C–C).

The area above the throat as it narrows toward the flue is called the smoke chamber, and this is where the smoke shelf is located, just behind the throat. The smoke shelf is necessary to prevent back drafts from blowing back down through the fireplace. The shelf (Figure 15.6) should be at least 8 in. wide or more, depending on the depth of the fireplace (over 20 in. should be 10 to 12 in.).

Figure 15.7. Back-to-back fireplaces.

Figure 15.8. Cross-sectional view of a typical fireplace. (U.S. Department of Agriculture.)

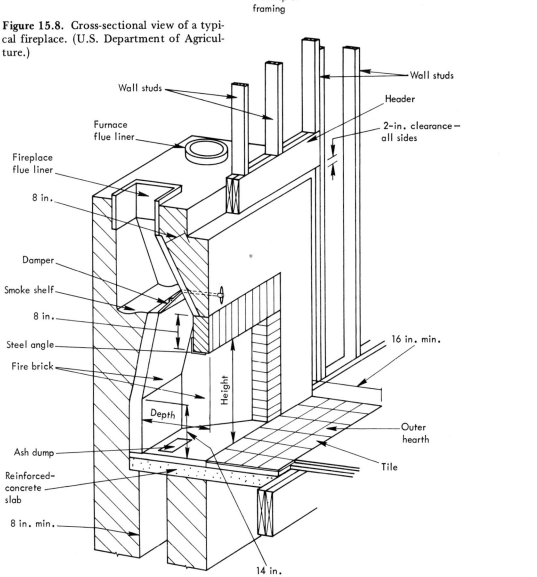

The damper is located at the throat and is of cast iron with a cast-iron frame being built into the masonry at the throat. The damper is necessary in severe climates to prevent cold air from entering the house, and a damper is useful in all climates. It regulates the flow of air up the chimney. It can be opened wide for a bright high fire and can be partially closed for a slow-burning fire. It can be closed completely in warm weather to prevent insects and rodents from entering the house.

Figure 15.7 shows two fireplaces back to back, and Figure 15.8 shows a cross section of a fireplace proper.

A fireplace that is well designed and well built is a source of satisfaction and a very definite sales asset.

15.2   STAIRS

Stairs are the basic means for the vertical movement of people and material. In building codes the stair is the only legal means of vertical transportation, no matter what other means are available. Stairs can be the feature of the entrance hall of a private residence. In the structure where there is adequate space and funds, the architect can plan an entrance hall around a good stairway. The builder or owner should build adequate and good-looking stairs. It is the first notable feature of the interior that a person sees. This first impression is valuable to a builder who may wish to sell the house; to an owner, who wishes to give a good impression to visitors, and it is a valuable resale tool.

15.2.1   Design as a Function of Use

Stairs must be designed so as to afford safe access between floors for people and materials. The stair should provide for easy ascent and descent and should provide ample headroom. The stairway should also add to the interior design. There are main stairs, service stairs, and basement stairs, and they are all designed for their special purpose. Although the main stair must be designed of first-grade materials and finish, the service stairs do not require such treatment, and basement stairs in most residences can be roughly built.

Types of Stairs

There are several types of stairways to serve their various purposes and to fit the space that is available. There is the straight run, which starts on one floor and runs straight up to the next floor. This straight run can be interrupted by a right-angled one- or two-step landing at the very top or bottom of the run. There can be a landing at midpoint to serve to turn the upper run at right angles to the lower run. There are also stairways with winders which serve to turn the stairway at a right angle without the use of a landing. Such winders should be avoided because the tread is pie-shaped and stepping on the narrow end of the tread can cause accidents. There are also stairways between the levels of split-level houses, winding stairways, disappearing stairs, and exterior stairs. Figure 15.9 shows a plan view of several of these stairways.

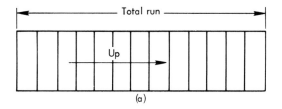

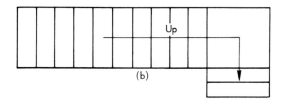

Figure 15.9. Types of stairs—plan: (a) A straight-run stair (note 14 risers); (b) landing turns stair at right angle (landing can be located any place in the run); (c) winders turn stair at right angle (note pie-shaped treads); (d) landing turns stair at 180° (suitable for small space).

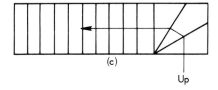

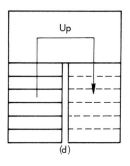

Main stairways should be built of good grades of hardwoods, such as oak, maple, and birch. The treads that get the hardest wear must be of first-grade hardwood, although the risers and stringers may be of a lesser grade or of a softwood. Basement stairs and service stairs may be of softwood, although the treads in service stairs should be of hardwood. Basement stairs can be made entirely of softwood, such as Douglas fir or southern pine.

15.2.2 Design for Convenience, Safety, and Maintenance

There are several basic minimum components of the design of stairs that must be adhered to if the desired function of a stair is to be attained. The word "minimum" is what building codes are based on and it is not the best quality. It is hoped that these "minimums" will be exceeded. The components are as follows:

Headroom

A primary consideration in stair design is sufficient headroom. For the main stairway the minimum headroom must be 6 ft 8 in. and for a service or basement stair it must be 6 ft 4 in.

Minimum Treads and Risers

For closed stairs (stairs with risers in place) a minimum requirement is a 9-in. tread and an 8¼-in. riser. A lower riser and a wider tread are more desirable. For instance, a good stair design would be a 10- to 10½-in. tread and a 7- to 7½-in. riser. The sum of riser and tread should fall between 17 and 17¾ in. Unfortunately, a stair with these comfortable dimensions may take too much space in a smaller house. It is to be noted that the minimum requirement is 9 + 8¼ = 17¼.

Ratio of Treads to Risers

This relationship has been established by long usage and is designed to give maximum safety and comfort to stair users. As mentioned above, the width of tread and riser when added should not exceed 17¾ in. or if multiplied by each other should be between 72 and 76 in. Example: 10½ by 7 = 73½, or 9 by 8¼ = 74¼, or 10 by 7½ = 75.

Handrails

These should be from 2 ft 6 in. to 2 ft 8 in. for runs and 2 ft 10 in. to 3 ft for landings.

Ratio of Floor-to-Floor Height to Length of Run

These ratios are shown in Table 15.4, together with other stair dimensions.

TABLE 15.4. Summary of Stair Dimensions

Floor-to-Floor Height	Total Run	Number of Risers	Riser (in.)	Tread (in.)	Height of Handrail
8'-0"	10'-3"	13	7⅜	10¼	2'-9"
8'-6"	11'-4¼"	14	7¼	10½	2'-9"
9'-0"	12'-3"	15	7¼	10½	2'-9"

The heights of the risers determine the number of steps between floors. If a floor to floor height of 8'-6" is used, there should be 14 risers (which is the commonly used number) of 7¼ in. and there will be a 10½-in. tread for a stair opening length or run of 11 ft 4¼ in. The clear width, as set by some codes, should be 3 ft 0 in., but many residential stairs have width of only 2 ft 8 in. This clear width means clear of the handrails. Less than 2 ft 8 in. will make it difficult to move furniture and can be considered somewhat of a hazard in case of fire. Figure 15.10 shows in line drawings how stair dimensions are determined.

15.2.3 Framing for Stairs

Chapter 9 showed how the builder must frame out an opening for a stair. The simplest way to frame a stair is when its run is parallel to the run of the floor joists, as shown in Figure 15.11. The joists on either side of the opening are doubled and should rest on a bearing wall. The ends of the opening are formed of double header joists.

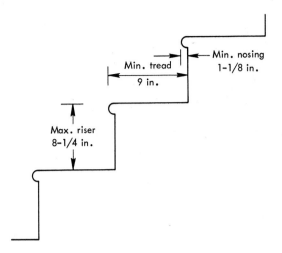

Min. tread
9 in.

Min. nosing
1-1/8 in.

Max. riser
8-1/4 in.

Figure 15.10.
Fundamentals in design of a stair.

Min. *H* main stairs 6' – 8"
Min. *H* service basement
stairs 6' – 4"

Parallel
lines

H

Fig. 15-10

Figure 15.11. Framing of
stair with run parallel to joists.

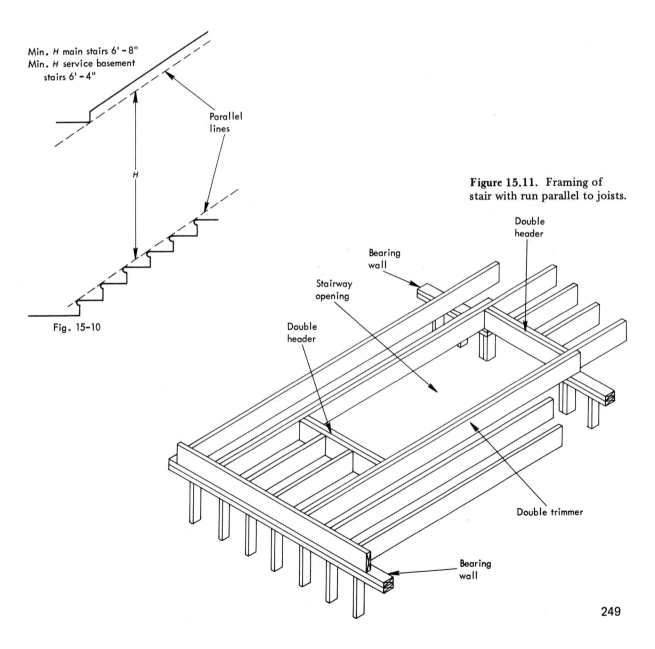

Double
header

Bearing
wall

Stairway
opening

Double
header

Double trimmer

Bearing
wall

It is much more difficult to frame as large an opening as is required for a stair when it is at right angles to the run of the joists. Figure 15.12 shows how this is done. It is to be noted that the joists on either side of the opening must support the weight of an entire floor system, which is carried by the joists that frame into it at right angles. These bearing joists (A) must, in turn, be supported by header joists (B). This requires heavy lumber and is not recommended if the framing shown in Figure 15.11 can be accomplished. However, if it must be done then such opening at right angle to the joist run should not exceed 10 ft in length. This can inhibit head room and a good stair.

A typical support for a landing which turns the stair at a right angle during its run is shown in Figure 15.13.

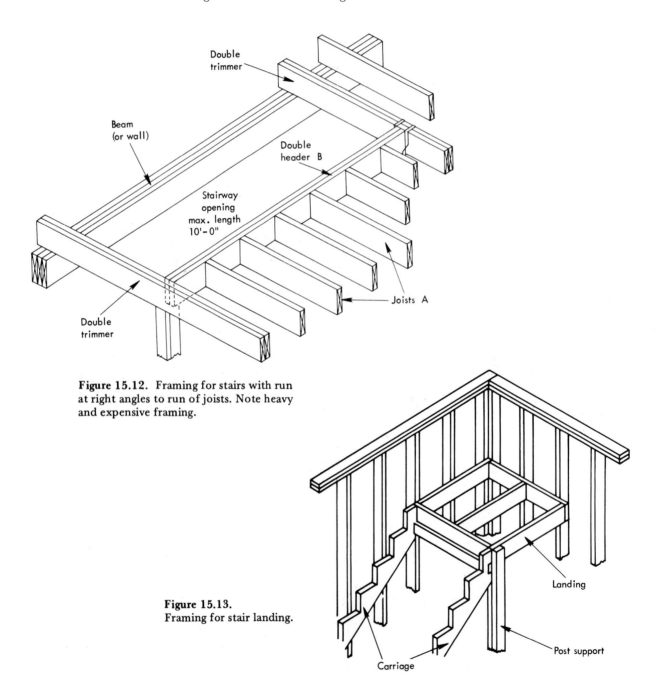

Figure 15.12. Framing for stairs with run at right angles to run of joists. Note heavy and expensive framing.

Figure 15.13. Framing for stair landing.

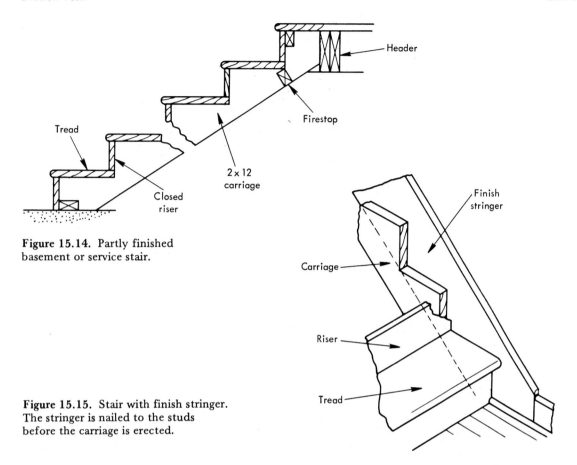

Figure 15.14. Partly finished
basement or service stair.

Figure 15.15. Stair with finish stringer.
The stringer is nailed to the studs
before the carriage is erected.

15.2.4 Stair Construction

It is important to have not only a well-designed and convenient stair but
also one that does not creak or have loose treads or otherwise cause a nuisance. The stairway must therefore be properly constructed. Most builders
have their stairs built by a professional stair builder who has the proper
tools and equipment, but many others build their own. In any case, every
builder or owner should know how a good stair is built.

The simplest stair to build is the basement stair, and this is usually built
by the builder. The stairway is supported by two planks or carriages, each
of which is usually a 2- by 12-in. plank of Douglas fir or southern pine.
These planks are cut to fit the predetermined width and height of tread and
risers. The carriages are supported by a double header at the top and are
braced against an anchor 2 x 4 at the bottom, as shown in Figure 15.14.
The stair as shown has risers in place and is a partly finished stair. A rough
basement stair has plank treads usually 1½ in. thick and open risers. A
third carriage is necessary if the stairway is wider than 3 ft unless the treads
are thicker than 1⅝ in.

The service stair should be of better appearance and quality than the
basement stair. It should have closed risers and a finish stringer as shown
in Figure 15.15. If the stair is enclosed between walls, the stringers should be
fastened to the walls before the rough carriages or the built stair is erected.

If one side is open, the finish stringer is fastened to this side of the stair after it is erected. The tread should be of good-quality hardwood.

As mentioned previously, the main stair is one of the noticeable and salable features of a house and the architect, owner, and builder should be willing to spend the effort and expense to do it well.

A main stair should be built of good hardwood. The supporting members should be housed stringers, as shown in Figure 15.16, instead of the rough carriages that can be used for other stairs. The stairs and treads should be fitted with routed and grooved joints, and the treads should have finished nosings, as shown in Figure 15.17. The open side or sides of the stair should be fitted with an ornamental baluster and newel post, as shown in Figure 15.18.

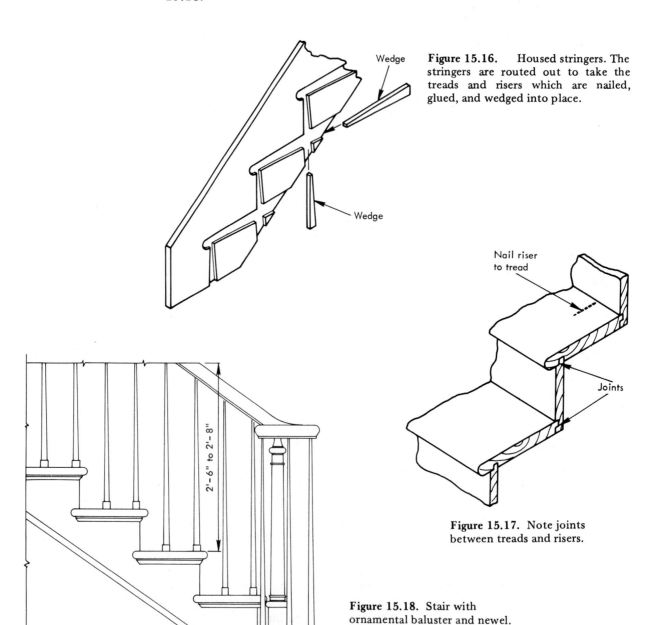

Figure 15.16.   Housed stringers. The stringers are routed out to take the treads and risers which are nailed, glued, and wedged into place.

Figure 15.17. Note joints between treads and risers.

Figure 15.18. Stair with ornamental baluster and newel.

Rough Opening	Ceiling Height	Stringers	Treads
25-1/2 x 54	7'5" to 8'9"	1 x 4	4 x 18
25-1/2 x 54	7'5" to 8'9"	1 x 5	6 x 18

Figure 15.19. Folding attic stairs and stock sizes.

15.2.5   Other Stairs

Attic Folding Stairs

Attic folding stairs are used only as a means of access to a storage space in the attic. Figure 15.19 shows how such a stair is installed. Such a stair requires very little clearance above, so that it can be used in a space under a low pitched roof. Two popular models have dovetailed treads into stringers, a spring mechanism to do most of the work in lowering and lifting the stair, and are made of clear vertical-grain lumber. The table beside the illustration gives framing and other dimensions.

Exterior Stairs

Exterior steps are used to come up to a front entrance or to a porch or patio. Very often (and advisably) such steps are made of masonry, which is not affected by weather, termites, rot, and other difficulties encountered by wood when it is used in contact with earth. If wood is used, it should rest on a masonry foundation and it should be carefully treated with wood preservative. Although the riser/tread ratio should be kept, the tread should be wider than indoors and the riser should be lower, such as: $10\frac{1}{2}$-in. tread and $6\frac{1}{2}$-in. riser, which totals 17 in.

An exterior stairway can also be used as the entrance and exit to a second-floor apartment with no other means of entry or exit. This stair then becomes a legal means of exit and is subject to code regulation.

CHAPTER SIXTEEN

finishing:
painting and decorating

16.1 FINISHING

The word "finishing," as its name implies, is the construction term for the various trades and processes that are involved in the final preparation of a structure for occupancy. (The actual specification term is "finishes.") Painting is almost one of the last of these processes. The finish coating of the interior and exterior surfaces protects them from weather, air pollutants, dirt and dust, and decay and also adds to the aesthetic effect of the structure.

16.2 EXTERIOR

Wood is one of the most extensively used materials in residential construction and in general is the material that more than almost any other requires the protection of a paint or a preservative to protect it against weather, decay, and infestation. The owner or builder should therefore pay particular attention to the kind of wood used for the exterior and to the kind of protective coating used to cover it. Painting the wood will, of course, totally obscure the grain or the natural color of the wood. The use of a preservative or a slightly colored stain will allow some of the grain and the wood color to show through. Both coatings are preventatives, but the builder must decide what architectural effect best suits the purpose. Many owners, in trying to preserve a naturally weathered appearance, will not coat shingles or siding, especially if they are of red cedar or redwood. If conditions are right, the wood will eventually turn gray and weather well for many years if it is not attacked by decay or insects.

16.2.1　Preservatives and Penetrating Stains

Instead of natural weathering, which has some elements of risk, the builder can use an almost transparent preservative which penetrates the wood and leaves a surface coating of water-repellant resin. This will prevent decay and mildew and still preserve most of the natural look. Another exterior finish is a penetrating oil stain that can be obtained as a semitransparent alkyd-oil stain or a fully pigmented acrylic latex stain (see Figure 16.1). Both these act as preservatives, with the semitransparent stain preserving some of the natural appearance of the grain of the wood.

Figure 16.1. Painted wood—exterior. (Courtesy Pittsburgh Paints, Pittsburgh, Pa. 15222)

painted wood – EXTERIOR

SPEC. NO.	USE: SURFACE AND REQUIREMENTS	TYPE FINISHING SYSTEM	SHEEN	RECOMMENDATIONS
1a	**WOOD SIDING** Fume resistant One coat white Extra high hiding Mildew resistant on paint film of 1-800 series.	Alkyd Chlorinated Paraffin, Linseed Oil	Gloss	Primer: *Speedhide* ® Exterior Wood Primer 6-9, or 6-809; or *Sun-Proof* ® 1 Coat Universal Primer 1-70, or 1-870. MWF 4.0 Mils Finish: *Sun-Proof* One Coat House Paint 1-45, or 1-845. 1 Coat White only. MWF 3.6 mils
1aa		Acrylic Latex	Flat	Primer: *Speedhide* Exterior Wood Primer 6-9, or 6-809; or *Sun-Proof* 1 Coat Universal Primer 1-70, or 1-870. MWF 4.0 mils Finish: *Sun-Proof* One Coat House Paint 70-45. White only. 2 Coats MWF 3.6 mils
1b	**WOOD SIDING ABOVE MASONRY** To avoid staining Slow chalking Fume resistant Good resistance to color fading) Mildew resistant on paint film of 1-800 series.	Oil Type Alkyd	Gloss	Primer: *Sun-Proof* Universal Primer 1-70, or 1-870; or *Speedhide* Exterior Wood Primer 6-9 or 6-809. 1 Coat MWF 4.0 mils Finish: *Sun-Proof* House and Trim Paint 1 line, or *Speedhide* Exterior 2 Coats Wood Finish 6 line. White or colors. MWF 3.2 mils
2	**WOOD SIDING** Properly prepared cement-asbestos and rustic wood shakes and shingles.	Acrylic Latex	Flat	Primer: *Sun-Proof* Latex House Paint Wood Primer 70-1. 1 Coat MWF 4.0 mils Finish: *Speedhide* Latex House Paint 6 line, or *Sun-Proof* Latex House 2 Coats Paint 70 line. White or colors. MWF 3.6 mils
2a			Gloss	Primer: (Same as 2 above) 1 Coat Finish: *Sun-Proof* Latex Gloss House and Trim Paint 78 line. White 2 Coats and colors. MWF 3.6 mils
3	**WOOD TRIM, SHUTTERS, DOORS** Accent areas, windows, handrails, etc. Not for use on large wood areas such as the body of a house.	Alkyd Enamel	Gloss	Primer: *Sun-Proof* Universal Primer 1-70, or 1-870; or *Speedhide* 1 Coat Exterior Wood Primer 6-9 or 6-809. MWF 4.0 mils Finish: *Sun-Proof* Oil Type House and Trim Color 1 line, or *Pittsburgh* 2 Coats Paints Quick Dry Enamel 54 line or 53 line, or *Speedhide* Exterior-Interior Enamel 6 line. White or colors. MWF 4.0 mils
3b	**WOOD TRIM, SHUTTERS DOORS, WOOD SIDING OR TRIM** (Large or small areas) Accent areas, windows, handrails, etc. only.	Acrylic Latex	Gloss	Primer: *Sun-Proof* Universal Primer 1-70, or 1-870; or *Sun-Proof* Latex 1 Coat Primer 70-1, or *Speedhide* Exterior Wood Primer 6-9 or 6-809. MWF 4.0 mils Finish: *Sun-Proof* Latex Gloss House and Trim Paint 78 line. White 2 Coats and colors. MWF 3.6 mils
4	**WOOD DECKS AND PORCHES** Tough, elastic film. Porches, decks, docks, floors, and wood steps. Resists abrasive underfoot wear	Alkyd Enamel	Gloss	Primer: *Pittsburgh* Paints Floor and Deck Enamel 3 line reduced with one 1 Coat pint mineral spirits or *Leptyne* ® Paint Thinner per gallon (for wood). For metal use proper inhibitive primers — *Speedhide* or PPG brands. Finish: *Pittsburgh* Paints Floor and Deck Enamel 3 line (full strength). 2 Coats White or colors. MWF 3.5 mils

16.2.2 Paints

Until fairly recently, one of the two most frequently used exterior paints was an oil-based paint that contained various proportions of titanium and zinc oxides and magnesium and calcium carbonates in a vehicle of linseed, safflower, or other oils. This paint has excellent covering quality and could be used over a single coat of primer or new wood. The other exterior paint is acrylic latex, which uses an acrylic resin emulsion as a vehicle for metallic oxides and carbonates and is thinned with water. This requires two coats over a coat of wood primer. In recent years, however, there has been a dramatic improvement in paint chemistry. Figure 16.1 includes examples of the types of exterior paint that are now available.

16.2.3 Preparation and Application

The preparation of the surface to be painted and the conditions under which the paint is applied is the most important part of the painting process. The final paint applied with the greatest skill will not produce a satisfactory finish unless the surface has been properly prepared. The surface must be free of dust or grease. Nail holes should be puttied and sanded. Any split shingle or siding should be repaired because weather can penetrate such cracks. Unless latex paint is used, there should be no painting on damp days. Painting is not recommended when the temperature is below $50°F$ or over $95°F$. Do not paint a cool, shaded surface if it will soon be exposed to hot sun. If the wet paint is dried too quickly, it may crack or blister. Try to follow the sun around the house.

The application of paint can be accomplished by brushing, the use of a roller, or the use of a paint sprayer.

Brush Painting

Painting with a brush is a familiar process. Certain tricks of the trade, however, will give a neater, more economical job. Dip the brush only about half the length to avoid buildup on the top of the bristles. The paint is laid on with short, vigorous strokes; then it is lightly cross-brushed with the tip of the brush to even it and eliminate brush strokes. The ceiling is painted first, then the walls, and the trim last. Corners and edges are painted so that the brush sweeps off the edge of the surface. The brush should not be poked into corners and edges; instead, touch these rough places lightly with the edge of the brush and if necessary twist the brush slightly.

Roller Painting

Rollers are being used more and more for painting, especially since such a wide variety has become available for different jobs. Rollers are made up of a cylinder which is attached to and rotates around a metal bar, which, in turn, is connected to a handle. Rollers range from $1\frac{1}{2}$ to 18 in. long and from $\frac{1}{2}$ to $2\frac{1}{2}$ in. in diameter. The fabric used to cover the cylinder varies with the type of paint used; the length of the fuzz or "nap" varies with the texture of the surface. Lamb's wool works well with oil-based paints, but it becomes matted if used with water-based paints. Nap lengths can range up to 1½ in., with the longer naps most useful for very rough surfaces. Mohair, made from the hair of Angora goats, comes with nap lengths of $\frac{3}{16}$ to ¼ in. It is especially good with enamels and on smooth surfaces. Two artificial fibers are used on rollers: dynel, which comes with nap lengths of ¼ to 1¼ in., is extremely good with latex paints; dacron, which is softer than dynel, is good with outdoor oil or water (latex) paints. Dacron comes in nap lengths ranging from $\frac{5}{16}$ to ½ in. Rollers are easy to use. The paint tray is filled about half full of paint. Then the roller is dipped into the paint until it is completely covered. Excess paint is squeezed out by pushing the roller up the ramp of the tray. The areas around the ceiling and the moldings are painted before the major parts of the surfaces are done.

Spray Painting

Spray painting utilizes a pump and an electric motor to force paint through a nozzle and onto a surface to be painted. In the most common paint sprayer, paint is poured into a closed container called a pot. Compressed air forces the paint through a hose to a spray gun. A separate hose is connected to the spray gun. The paint is then atomized by air blown through an opening in the caps of the spray gun. There are several things to keep in mind when setting up and using spray equipment. The paint must be thin enough to be atomized and turned into a spray but thick enough to stick to a surface without running or dripping. The air pressure on the pot of paint determines exactly how much paint will flow into the nozzle. If too much air is fed in, the paint will atomize too much and a fine, dry spray will be produced. If too little pressure is applied, the paint spatters and leaves a speckled, almost gritty, coat. The spray gun should be held 6 to 10 in. away from the surface to be painted. If it is held much farther away, the spray will "dust"; that is, the droplets will dry up in midair and hit the surface as hard, solid particles. The spray gun is kept at right angles to the surface to be covered. The gun is held straight and not tilted. The painter's arm moves back and forth parallel to the surface. The spray pattern should feather. This can be done if the painter will pull the trigger on the gun after the stroke begins and release it before it ends. On corners, the paint stops 1 to 2 in. before the corner. Then the gun is swept up and down the corner so that both sides are hit. Figure 16.2 shows how a sprayer should be used.

Masonry

Brick, stucco, and concrete walls can be coated with the paints as shown in the Figure 16.3. The painting of masonry can be subject to variations as described by the following extract from an article in a U.S. government booklet.

Masonry surfaces which are frequently painted are concrete, stucco, and plaster. They have several things in common: they are hard, contain lime, and are fairly porous. New masonry may be either rough or smooth. Smooth concrete in particular is often difficult to paint because the paint will flake and peel off. It will not stick to a smooth surface. Another problem with fresh mortar is that the lime makes it "alkaline." Alkaline substances will turn oil-based paints into a greasy soap-like substance which will not stay on a wall. Still another problem which occurs with masonry surfaces is called efflorescence. Concrete, stucco, and plaster all contain salts which dissolve in water. If water soaks through the masonry it will dissolve these salts and they will crystallize on the surface of the masonry. If the paint is water soluble, these salts will soak through the paint and crystallize on the surface, creating unsightly spots. If the paint is not soluble, these crystals will form beneath the surface of the paint and create blisters. Either way, efflorescence has to be washed away and its cause removed before a masonry surface can be painted. Differences in quality may occur in different spots in the masonry. If the plaster, stucco, or cement was mixed, applied, or cured improperly, different parts of the surface may absorb different amounts of paint. This means that a coat of paint will be glossier on some parts of the wall than on others. This sort of surface will cause paint to flake easily and results in a poor-quality paint job.

Figure 16.2. Technique for use of paint sprayer.

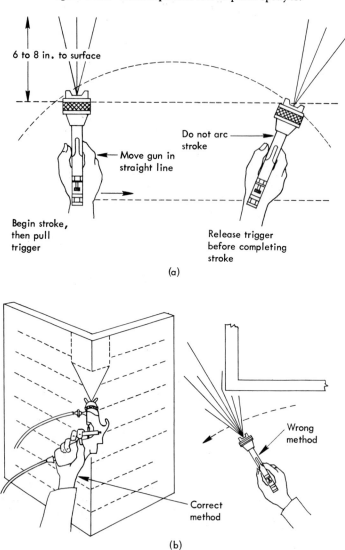

6 to 8 in. to surface

Do not arc
stroke

Move gun in
straight line

Begin stroke,
then pull
trigger

Release trigger
before completing
stroke

(a)

Wrong
method

Correct
method

(b)

10	**BRICK, STUCCO, AND CONCRETE WALLS:** Durable, attractive, and weather-resistant film for use over surfaces free of chalky deposits and efflorescence.	Vinyl Acrylic Latex	Flat	**Primer:** None required. (Or use 6-7 Latex Block Filler for smoother finish.) **Finish:** *Cementhide* ® Latex Masonry Paint 37 line or *Speedhide* Emulsion **2 Coats** Masonry Paint 6 line. (Reduce first coat with one pint water per gallon for porous surfaces.) White and Colors. MWF 5.4 mils
10a	**BRICK, STUCCO, AND CONCRETE WALLS:** Durable, attractive, and weather-resistant film for use over surfaces free of chalky deposits and efflorescence.	Acrylic Latex	Flat	**Primer:** PPG Alkali Resistant Primer 6-3. **1 Coat** **Finish:** *Sun-Proof* Latex House Paint 70 line, or *Speedhide* Exterior Latex **2 Coats** Wood Finish 6 line. White and Colors. MWF 4.6 mils **or** *Sun-Proof* One Coat Latex House Paint 70-45. White only. **1 Coat** MWF 3.6 mils
10c			Semi-Gloss	**Primer:** PPG Alkali Resistant Primer 6-3. **1 Coat** **Finish:** *Sun-Proof* Latex Semi-Gloss and Trim House Paint 78 line. White **2 Coats** and Colors. MWF 3.6 mils
13	**CEMENT-ASBESTOS SIDING, AND PANELS:** Note: — These specifications do not apply to the sandwich-type fiber core insulation boards.	Vinyl Acrylic Latex	Flat	**Primer:** PPG Alkali Resistant Primer 6-3. **1 Coat** **Finish:** *Cementhide* Latex Masonry Paint 37 line, or *Speedhide* Emulsion **2 Coats** Masonry Paint 6 line. White and Colors. MWF 5.4 mils
14	**CONCRETE PATIO AND BREEZEWAY FLOORS** Durable finish capable of withstanding normal washing but not prolonged soaking.	Urethane-Latex	Eggshell	**Primer:** *Pittsburgh* Urethane-Latex Floor, Deck and Patio Enamel 3 line **1 Coat** reduced with one pint water per gallon. Acid etching is not required but will improve adhesion to smooth dense concrete. **Finish:** *Pittsburgh* Urethane-Latex Floor, Deck and Patio Enamel 3 line. **2 Coats** White and Colors (After curing for several days, apply a quality Paste Wax.)
14a	**FOR NEW OR PREVIOUSLY PAINTED CONCRETE FLOORS** Durable, tough coating with good hiding qualities. Also suitable for stucco. Interior and exterior use.	Urethane-Latex	Eggshell	**Primer:** Thin with 1 pint of water per gallon on new, unpainted surfaces. **Finish:** PPG Urethane-Latex Floor, Deck and Patio Enamel 3 line in **1 or 2** selected color. 2 coats preferred, MWF 3.6 mils per coat. **Coats**
14b		Alkyd-Oil	Gloss	**Primer:** (After acid etching) PPG Alkali Resistant Primer 6-3. **1 Coat** **Finish:** *Pittsburgh* Alkyd Floor and Deck Enamel 3 line. **2 Coats** White and Colors.
14d	**EXTERIOR CONCRETE ROADS, DRIVEWAYS, ETC.**	Alkyd and Chlorinated Rubber	Flat	**Primer:** None needed. **Finish:** *Pittsburgh* Traffic and Zone Marking Paint 11 line. # **2 Coats** MWF 5.0 mils
14e	**NEW EXTERIOR MASONRY SURFACES** Concrete, Block, Brick, Stucco. Prime, fill and finish in one application, using airless spray	Vinyl Acrylic Latex	Semi-Gloss	**Primer:** None usually required EXCEPT chalky, weathered exterior surfaces must be sealed prior to application. Use PPG Masonry Sealer 6-8 (37-40). **Finish:** One coat is usually sufficient using *Speedhide* Prime, Fill and **One or** Finish 6-507 in airless spray. After spraying, rollers can be **Two** used to release trapped air and minimize pinholing. On **Coats** exterior applications if pinholes or cracks are noticed, a second coat may be used to insure a good weather seal.

Figure 16.3. Masonry—exterior. (Courtesy Pittsburgh Paints, Pittsburgh, Pa. 15222)

Metals

Galvanized steel surfaces, such as gutters and downspouts, should be primed with recommended special primers, since conventional primers usually do not adhere well to this type of metal. A zinc-dust-oxide-type primer works well on galvanized steel. Exterior latex paints are sometimes used directly over galvanized surfaces, but oil paints are not so used.

Unpainted iron and steel surfaces rust when exposed to the weather. Rust, dirt, oils, and old loose paint should be removed from these surfaces by wire brushing or power tool cleaning. The surface should then be treated with an anticorrosive primer.

Zinc-chromate-type primers are effective on copper, aluminum, and steel surfaces, but other types are also available for use on metal.

16.3 INTERIOR

16.3.1 Woodwork and Trim

Interior finishing should be more oriented toward the excellence and appearance of the surface than toward protection against weather or decay. For interior woodwork there are transparent and opaque finishes. The transparent finishes are used if the natural color and grain of the wood is to be preserved. The wood is first coated with a thinned-out lacquer sealer, which is usually made of a thin solution of pale shellac. The wood is then coated with gloss or semigloss varnish or wax. The glossy coatings can be dulled down with pumice and the wax can be rubbed to a sheen.

Opaque finishing of interior woodwork is usually accomplished by the use of a gloss or semigloss enamel. Such paint presents a harder finish to withstand moisture and washing, especially in such spaces as kitchen and bathrooms. Before the application of enamel, the wood surface should be perfectly smooth and perfectly dry. Knot holes should be painted over with knot sealer. Nail holes should be puttied and sanded. The builder should allow the house to become thoroughly dry in summer or winter before performing any interior painting, especially in the use of enamels. Figure 16.4 gives a list of the paints that are available for interior woodwork and trim.

16.3.2 Walls

The great majority of interior walls in residential construction are constructed of dry wall (wallboard). Chapter 12 has described the proper way of installing wallboard. The painter must look over such walls very carefully to ascertain that all joints have been properly sealed and sanded and that all nailholes or other imperfections in the surface have been properly covered and smoothed. If any cracks have developed, they must be cut out and spackled with spackling compound, then sanded smooth.

Walls can be coated with oil-based or water-based (latex) paint. Figure 16.5 shows a choice of the paints that are available for interior walls, whether drywall or plaster. Oil-based paint can be applied only to thoroughly dry plaster. At least 1 month of drying is recommended. If there is any doubt, the use of a moisture meter (available at most paint stores) is advised.

16.3.3 Floors

The use of wood for flooring in residential structures is a very desirable finish. Wood is the most durable of residential floor finishes and is by far the most attractive. The finishing of a wood floor should be such as to best bring out the natural grain of the wood that is used and to generally enhance the appearance of the room.

The finishing of floors consists of several different operations. First the floor must be sanded, and the sanding should be done by a machine that is used by an experienced operator to make sure that it is evenly done with no grooving or other marking. The sanding must progress from rough to fine paper (2/0) to provide a very smooth surface. If the wood is not very close grained, a paste or liquid filler is then used to fill in all open pores.

painted woodwork & trim – INTERIOR

SPEC. NO.	USE: SURFACE AND REQUIREMENTS	TYPE FINISHING SYSTEM	SHEEN	RECOMMENDATIONS
37	**WOODWORK AND TRIM** (Painted) Resists yellowing. Very high hiding power with exceptional opacity and filling properties. Smooth, brilliant finish is easy to maintain.	Solvent Type Alkyd Enamel	Gloss	**Primer:** PPG Quick-Dry Enamel Undercoater 6-6 (54-255), reduced **1 Coat** slightly with *Leptyne* paint thinner or mineral spirits. Or use PPG Water Base Enamel Undercoater 6-755. MWF 3.5 mils **Finish:** *Pittsburgh* Paints Quick-Dry Architectural Gloss White **1 or 2** Enamel 54-352. **Coats** May be tinted. MWF 3.5 mils
37a		Water Base Alkyd Enamel	Gloss	**Primer:** PPG Quick-Dry Enamel White Undercoater 6-6 (54-255) **1 Coat** reduced slightly with *Leptyne* paint thinner or mineral spirits; OR use PPG Water Base Enamel Undercoater 6-755. MWF 3.5 mils **Finish:** PPG 53 line or *Speedhide* Water Base Gloss Enamel 6 line. **1 or 2** White and Colors. **Coats** MWF 3.5 mils
38	**WOODWORK AND TRIM** (Painted) White. Resists yellowing. Very high hiding power. Applicable to interior surfaces where fine rubbed-effect finish is desired.	Alkyd Enamel	Eggshell Flat or Rubbed Effect (Subdued)	**Primer:** PPG Quick-Dry Enamel Undercoater 6-6 (54-255) reduced **1 Coat** slightly with *Leptyne* paint thinner or mineral spirits; OR use PPG Water Base Enamel Undercoater 6-755. MWF 3.5 mils **Finish:** *Pittsburgh* Paints Quick-Dry Architectural Eggshell White **1 or 2** Enamel 54-362. May be tinted. **Coats** MWF 3.5 mils
39	**WOODWORK AND TRIM** (Painted) High gloss finish enamel that resists chalking, yellowing and color fading. Applicable to exterior as well as interior surfaces.	Alkyd Enamel (Solvent Type)	Gloss	**Primer:** PPG Quick-Dry Enamel Undercoater 6-6 (54-255), reduced **1 Coat** slightly with *Leptyne* paint thinner or mineral spirits; OR use PPG Water Base Enamel Undercoater 6-755. MWF 3.5 mils **Finish:** *Pittsburgh* Paints Quick-Dry Enamel 54 line or *Speedhide* **1 or 2** Exterior-Interior Gloss Enamel 6 line. White and Colors. **Coats** MWF 3.5 mils
39a		Alkyd Enamel (Water Base Type)	Gloss	**Primer:** PPG Quick-Dry Enamel Undercoater 6-6 (54-255) reduced **1 Coat** slightly with *Leptyne* paint thinner or mineral spirits. Or use PPG Water Base Enamel Undercoater 6-755. MWF 3.5 mils **Finish:** PPG 53 line or *Speedhide* Water Base Gloss Enamel 6 line. **1 or 2** White and Colors. **Coats** MWF 3.5 mils
40	**WOODWORK AND TRIM** (Painted) Resists yellowing. Durable finish with exceptional hiding power. Easy to apply over walls as well as trim. Easy to maintain.	Alkyd Enamel	Satin	**Primer:** PPG Quick-Dry Enamel Undercoater 6-6 (54-255) or **1 Coat** PPG Water Base White Undercoater 6-755. MWF 3.5 mils **Finish:** *Satinhide* Alkyd Lo-Lustre Enamel 20 line or *Speedhide* **1 or 2** Alkyd Interior Lo-Sheen Enamel 6 line. White and Colors. **Coats** MWF 3.6 mils
40a		Acrylic Latex	Semi-Gloss	**Primer:** PPG Quick-Dry Enamel Undercoater 6-6 (54-255) or **1 Coat** PPG Water Base White Undercoater 6-755. MWF 3.5 mils **Finish:** *Satinhide* Latex Lo-Lustre Enamel 88 line or *Speedhide* **1 or 2** Latex Semi-Gloss Enamel 6 line. White and Colors. **Coats** MWF 3.6 mils
40b	**WOODWORK AND TRIM** (Painted) Latex Enamel, high resistance to soil. Excellent washability. Wide color selection. Also suitable for trim, cabinets, other objects and surfaces.	Acrylic Latex	Eggshell Flat or Semi-Gloss	**Primer:** PPG Quick-Drying Enamel Undercoater 6-6 (54-255) or PPG **1 Coat** Water Base White Undercoater 6-755. MWF 3.5 mils **Finish:** *Manor Hall* Latex Flat Enamel 89 line or *Manor Hall* **1 or 2** Lo-Lustre Enamel Designer Bases 89 line. White and Colors. **Coats** MWF 3.6 mils
41a	**WOODWORK AND TRIM** Extremely durable finish for door frames and trim where repeated washings are needed or stains may occur.	Two-Component Polyester-Epoxy or Acrylic Epoxy	High Gloss, or Semi-Gloss	**Primer:** PPG Quick-Drying Enamel Undercoater 6-6 (54-255) or **1 Coat** PPG Water Base White Undercoater 6-755. MWF 3.5 mils **Finish:** *Pitt-Glaze* High Solids Polyester-epoxy Solvent Base or **1 or 2** High Solids Acrylic-epoxy Water Base Coating, 16 line in **Coats** finish and color desired. White and Colors. MWF 6 mils — Solvent Type MWF 4 mils — Water-Base Type
41b	**WOODWORK, TRIM, OR WOOD PANELING** To retard flame spread and for short term insulation of structural members in case of fire.	Modified PVA Latex Intumescent Fire Retardant	Flat	**Primer:** (Solvent base) PPG Quick-Drying Enamel Undercoater. **1 Coat** 6-6 (54-255). MWF 3.5 mils OR (Water Base) PPG Quick-Dry Emulsion Sealer 6-2 (17-10) MWF 4.0 mils **Finish:** *Speedhide* Latex Fire Retardant Paint 42-7. White and Colors. **2 Coats** MWF 5.3 mils

Figure 16.4. Painted woodwork and trim—interior. (Courtesy Pittsburgh Paints, Pittsburgh, Pa. 15222)

Such woods as oak, walnut, and mahogany require a filler. The finish can be a sealer, which barely penetrates the wood and which allows a final coating of paste wax, or it can be a varnish based on resin, such as alkyd, polyurethane, or epoxy. Any of these latter provide a hard, lustrous finish. The use of stain to bring out a particular grain or color is at the choice of the builder or owner. It should be oil-based and should be used after the filler. Stain is not generally used in modern houses.

Figure 16.5. Dry wall—interior.
(Courtesy Pittsburgh Paints, Pittsburgh, Pa. 15222)

dry wall -INTERIOR

32	**DRY WALL CONSTRUCTION:** Low odor, resists yellowing. For use over interior plaster, wallboard, and wood. Paper-type surfaces must be sealed with an emulsion type sealer before applying finish.	Alkyd-Resin	Flat	Primer: PPG Quick-Drying Emulsion Sealer 6-2 (17-10). 1 Coat MWF 4.6 mils Finish: *Speedhide* 6 line or *Wallhide* Alkyd Flat 24 line. White and 1 Coat Colors. MWF 4.6 mils
32a	**DRY WALL CONSTRUCTION:** Enamel, high resistance to soil and washing. Wide color selection. Also suitable for trim, cabinets, other objects and surfaces.	Acrylic Latex	Eggshell Flat,	Primer: *Manor Hall* Latex Flat Enamel 89 line reduced with one pint of 1 Coat water per gallon; or use PPG Quick-Drying Emulsion Sealer 6-2 (17-10). MWF 3.6 mils. Finish: *Manor Hall* Latex Flat Enamel 89 line. White and Colors. 1 or 2 MWF 3.6 mils Coats
32b	**DRY WALL CONSTRUCTION** Low odor, resists yellowing. For use over interior plaster wallboard, and wood. Paper-type surfaces must be sealed before applying this finish.	Alkyd Resin	Semi-Gloss	Primer: PPG Quick-Drying Emulsion Sealer 6-2 (17-10). 1 Coat MWF 4.6 mils Finish: *Speedhide* Alkyd Lo-Sheen Enamel, 6 line; or *Satinhide* 1 Coat Lo-Lustre Alkyd Enamel 20 line. White and Colors. MWF 3.6 mils
32c	**DRY WALL CONSTRUCTION** Latex Enamel, high resistance to soil and washing. Wide color selection. Also suitable for trim, cabinets, other objects and surfaces.	Acrylic Latex	Semi-Gloss	Primer: PPG Quick-Drying Emulsion Sealer 6-2 (17-10). 1 Coat MWF 3.6 mils Finish: *Speedhide* Latex Interior Semi-Gloss Enamel, 6 line; or 1 or 2 *Satinhide* Lo-Lustre Latex Enamel 88 line. Coats White and Colors. MWF 3.6 mils
35	**DRY WALL CONSTRUCTION** Applicable also to plaster, wallboard, cement-masonry and brick.	Vinyl Acrylic or Acrylic Latex	Flat	Primer: 2 Coats *Speedhide* Acrylic Latex Interior Flat Wall Paint 6 and line, or One or Two Coats of *Wallhide* Vinyl Acrylic Latex Finish: Flat Wall Paint 80 line. White and Colors. MWF 4.0 mils
35a	**DRY WALL CONSTRUCTION** Applicable also to plaster, wallboard, cement-masonry and brick. Exceptional hiding power over interior surfaces. Dries to handling in 30 minutes to one hour. **For spray application only.**	Spray-applied Vinyl Acrylic Latex	Velvet Flat or Textured	Primer: No primer needed on wall board or plaster. Finish: *Hide-A-Spray* High Build Latex Flat Paint 91 line. White and 1 Coat Bone White. White may be tinted. Normal application is 6 to 10 mils wet. (Textured finishes also available.)
36	**DRY WALL CONSTRUCTION** Extremely hard-wearing surface treatment for institutional or industrial-commercial areas that require cleaning with strong detergents and mechanical scrubbers. Provides sanitary surface.	Two-Component Polyester-Epoxy or Acrylic Epoxy	High Gloss, or Semi-Gloss	Primer: *Pitt-Glaze* Pigmented Sealer 16-8. 1 Coat MWF 3.2 mils Finish: *Pitt-Glaze* High Solids Polyester-epoxy Solvent-Base, or High 1 or 2 Solids Acrylic-epoxy Water-base, 16 line, Gloss or Semi-Gloss Coats Coating. White and Colors. MWF 6 mils — Solvent Type MWF 4 mils — Water Base
36a	**DRY WALL CONSTRUCTION** To retard flame spread and for short term insulation of structural members in case of fire.	Modified PVA Latex Intumescent Fire Retardant	Flat	Primer: PPG Quick-Drying Emulsion Sealer 6-2 (17-10) or PPG 1 Coat Quick-Dry White Undercoater 6-6 (54-255). MWF 4.0 mils Finish: *Speedhide* Latex Fire Retardant Paint 42-7. 2 Coats MWF 5.3 mils

16.4   DECORATING

Painting of walls, floors, and trim decorates these surfaces because it is used to enhance their appearance. There are, however, wall finishes other than paint.

16.4.1   Wallpaper and Fabrics

The most important of such finishes are wallpaper and fabrics. These finishes are not generally used in the ordinary run of residential construction. For the owner or builder who wishes to use them and is willing to pay the extra cost entailed, there is a wide variety of wallpaper and fabrics available which can be used for almost any decorative purpose. They range from scenic wallpaper to grass cloth to flocked paper to vinyls for use in bathrooms and kitchens. Most modern wallpapers are made of a vinyl-treated fabric rather than of paper. Many wallpapers are now backed with paste ready to be wetted and hung.

The most important precaution in the application of any wall covering is that the surface be smooth and dry. All nail holes and cracks must be filled and sanded. It is recommended that new walls be coated with a glue size, which seals the pores and presents an even surface for the adherence of the wall covering.

CHAPTER SEVENTEEN

plumbing, heating, air conditioning, and electrical systems

This chapter will not concern itself with how to install the mechanical and electrical systems in a structure. It will contain an outline description of these systems so that the owner or builder can be aware of the essential characteristics of a good system. All of these trades require a high degree of skill and experience and should not be lightly undertaken.

The importance of these trades to safety and health is underlined by the building codes. In many states and municipalities an owner or builder is forbidden from installing any of these systems unless he or she has a license (which is usually obtained by examination). The reasons for this strict attitude on the part of the authorities can be simply explained.

An improperly installed plumbing and drainage system can contaminate the potable water supply and can create a possible health problem. Typhoid, dysentery, and other illnesses are usually caused by a contaminated water supply or a poor drainage system.

An improperly installed heating system can be the cause of discomfort and a waste of fuel. In forced-air systems the leakage of furnace gases can be a serious health menace.

An improperly installed oil burner can be a fire hazard.

A poorly installed electrical system can present a fire hazard and if not properly grounded can cause electrocution.

These items are mentioned not to frighten anyone but simply to emphasize the problems and the knowledge and skill that are involved.

17.1 PLUMBING

The plumbing system of a building consists of a potable water supply, sanitary facilities, sanitary and storm drainage, and sanitary storm water and sewage disposal.

Even in communities where an owner is allowed to perform his or her own plumbing work, the owner has to file a plan and obtain a permit and the work is subject to inspection by the local building inspector.

17.1.1 Minimum Requirements

Following are some of the important plumbing provisions that are enforceable by law.

All premises intended for human habitation shall be provided with an adequate supply of pure and wholesome potable water which must not be connected to any unsafe water supply or subject to backflow or back siphonage.

All plumbing fixtures in any building that abut on a street or an easement connected to a street that contains a sewer must connect to the sewer. If there is no sewer, a suitable disposal system must be approved by local authority. (*Note:* The builder or owner is strongly advised to consult such an authority before final costs are determined. Disposal systems can be very expensive and very troublesome if not correctly installed.)

Every dwelling unit must have at least one water closet, one tub or shower, one lavatory, and one kitchen sink. The fixtures must be of nonabsorbent materials and must have water seal traps.

Every plumbing system must have a venting system that vents to the outer air and sanitary cleanouts. Where necessary, they must also have backwater traps.

17.1.2 Plumbing Materials

Piping

The use of copper tubing for hot- and cold-water supply lines is almost universal in all parts of the country in which "soft" water is used. Red brass has also been used, but the soldered-joint, drawn, and annealed copper tube has been found to be the most economical and longest lasting. Galvanized wrought iron or steel piping can be used in parts of the country where "hard" water is prevalent. With this type of water the piping is not subject to corrosion and is quite satisfactory.

Plastic pipe for drainage lines and water lines is now allowed by many plumbing codes and is being used almost exclusively where it is allowed. It is impervious to many corrosive materials, it is light in weight, it is easy to cut, and if properly connected, the joints are stronger than the pipe itself. Plastic piping must be well supported. Sewage lines can be plastic inside the house, but most codes require cast iron for the underground exterior line.

showers
fiberglass-reinforced polyester

Master-size, luxury one-piece showers with restful seats for grooming and relaxation.

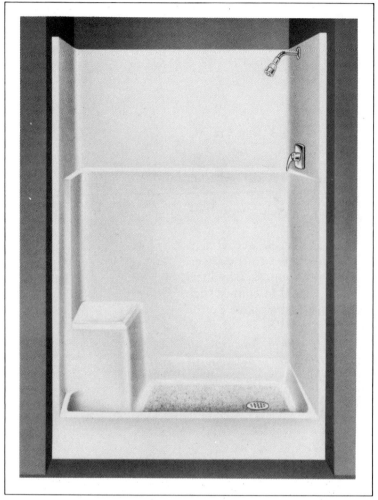

Shown with 1490.051 Aquarian II shower fitting. 3570.017 C.P. 4" (102mm) shower drain. (Drain must be specified separately. Shower not drilled for other fittings.)

Fiberglass-reinforced polyester. (Courtesy American Standard)

Integral shower/wall surround

Designer Line recess integral shower/wall surround with optional ceiling—outlet on right or left—soap ledge—nailing flange—fiberglass-reinforced polyester—seat and slip-resistant bottom.
(drain not included with unit)

(shower only)
Model 148
36" x 48" (915mm x 1219mm)
2156.016 right outlet *(illustrated)*
2156.032 left outlet
Model 160
36" x 60" (915mm x 1524mm)
2156.024 right outlet
2156.040 left outlet

Model 148
2156.016/.032
Model 160
2156.024/.040
a—36" (914mm)
b—73½" (1867mm)
c—48" (1219mm)
d—60" (1524mm)
e—62½" (1588mm)
f—2" (51mm)
g—6" (152mm)
h—18" (457mm)
i—13" (330mm)
j—1½" (38mm)

Model 148
2156.073 standard ceiling
Model 160
2156.099 standard ceiling
a—47⅞" (1216mm)
b—59⅞" (1521mm)
c—13½" (343mm)
d—35⁷⁄₁₆" (900mm)

Shower door opening
45¾" (1162mm) wide
57¾" (1467mm) wide

Fixtures

Bathtubs are available in fiberglass-reinforced polyester, enameled cast iron, and enameled steel, in that order of cost. The enameled sheet steel is the least expensive. The builder is advised to pack rockwool under this tub to deaden its noise characteristics. Preformed fiberglass tub and shower combinations, as well as shower enclosures, are now available, as shown in Figures 17.1 and 17.2.

Lavatories are available in vitreous china, which is by far the best-looking and best-wearing material. There is also enameled cast iron and Duramel, as well as various cast plastics. Lavatories come with fitted countertops and with undersink cabinets, which are almost a necessity in a modern house.

Water closets come only in vitreous china. One-piece units with the flush box as part of the toilet bowl are very handsome and expensive. Close-coupled units are less expensive, and there are units in which the flush box and the bowl are connected only by piping.

Kitchen sinks are available in stainless steel and in enameled cast iron.

The builder or owner has available a wide choice of plumbing fixtures as to material, appearance, and price range. It is recommended that such fixtures be the best that can be afforded. Plumbing fixtures last a long time; good quality is trouble-free and certainly makes a house more salable.

The builder or owner should also make sure that the plumbing contractor installs water supply and drainage lines of sufficient size so that there will be a good supply of water to each water outlet when other outlets are in use. There are tables available which tell how many gallons per minute should be available for each fixture and the size of pipe that will accomplish this. (Copper Development Association, Inc., 1011 High Ridge Road, Stamford, Conn. 06905.) Consult the local building officials, who will be familiar with minimum code requirements for supply and drain-line sizes.

17.2 HEATING

17.2.1 Heating Systems

In cold climates where central heating is required, the builder has a number of choices regarding the kind of system to install and the kind of fuel to use. The two heating systems most widely used are forced warm air and forced hot water.

The forced-warm-air system can be adapted to be used for both heating and air conditioning. The air is heated (or cooled) and filtered, and a blower then forces the air to the various spaces through ducts ending in registers. Return air is collected at one or more central spots or in each room, from which a duct carries it back to the furnace.

Figure 17.3 shows a simple forced-air system. The system is quick to respond to changes in temperature and is probably the most reasonable to install. It is almost a necessity in regions where central air conditioning is a requirement. The system is also used in houses on concrete slabs. Figure 17.4 shows a forced-air perimeter-loop system. The heating plant itself can be quite small and can be placed in an attic or in a well-ventilated closet.

Designer Line Bath

AMERICAN STANDARD

Designer Line recess integral bath/wall surround—5′ length—nailing flange—fiberglass reinforced polyester

catalog number (tub and wall surround only)

Model 170—14″ apron
☐ **2146.033** right outlet
☐ **2146.041** left outlet

for above floor drain installation
Model 270—16⅛″ apron
☐ **2146.058** right outlet
☐ **2146.066** left outlet

Model 370—20″ apron
☐ **2146.074** right outlet
☐ **2146.082** left outlet

nominal dimensions

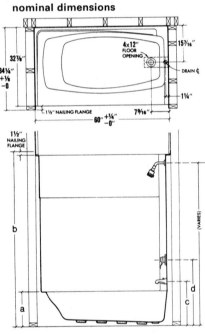

Catalog number	a	b	c	d
2146.033/2146.041	14″	73½″	18″	32″
2146.058/2146.066	16⅛″	75⅝″	19⅛″	34⅛″
2146.074/2146.082	20″	79½″	24″	38″

Color:

(a)

Figure 17.2. Integral bath and shower: Fiberglass-reinforced polyester.
(a) Designer line bath.
(b) Designer line shower.
(Courtesy American Standard)

Designer Line shower

Designer Line recess integral shower/wall surround—outlet on right or left—nailing flange—fiberglass reinforced polyester—with drain—slip-resistant bottom and seat

catalog number

Model 148—36 x 48″
☐ **2156.016** right outlet
☐ **2156.032** left outlet

Model 160—36 x 60″
☐ **2156.024** right outlet
☐ **2156.040** left outlet

nominal dimensions

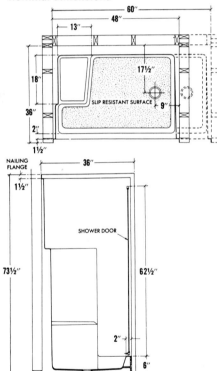

4″ grid drain with 2″ caulked outlet furnished with unit

Color:

(b)

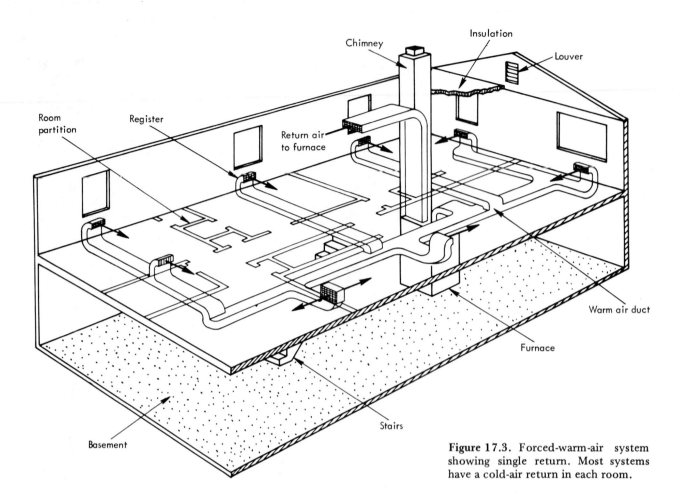

Room
partition

Register

Chimney

Insulation

Louver

Return air
to furnace

Warm air duct

Furnace

Stairs

Basement

Figure 17.3. Forced-warm-air system showing single return. Most systems have a cold-air return in each room.

The most frequently used hot-water systems are the series-loop system, which simply conducts the pumped hot water from one radiator to the next in series and finally returns the water to the boiler, and the one-pipe system, which by means of a directional valve sends the hot water into and through a radiator and then returns it to the one-pipe loop. Figures 17.5 and 17.6 show these systems. Because of the bulk of the metal that is heated in the piping and radiators and which loses heat slowly, hot-water heat is very even and not subject to the quick changes in temperature that occurs in a warm-air system. It is an excellent heating system to use if no central air conditioning is contemplated. In larger homes a forced-hot-water system can also be very easily zoned so that a separate loop going to an upper floor and controlled by a small circulating pump and thermostat can keep bedrooms at a lower temperature than the first-floor living quarters—which is very important in these days of high energy costs—and is an excellent sales point.

Figure 17.7 shows an oil-burner installation for a forced-hot-water system. This shows the intricacies of such an installation, which should be done only by a licensed installer. Figure 17.8 shows baseboard radiators as used in a forced-hot-water system. The rating data show capacity Btu per hour per lineal foot of radiator at various water temperatures.

In warm climates, where heat is required only on a standby basis, an oil-fired or natural-gas-fired underfloor furnace or electrical-resistance-heated wall panels can be used. This does not, of course, provide for central air conditioning.

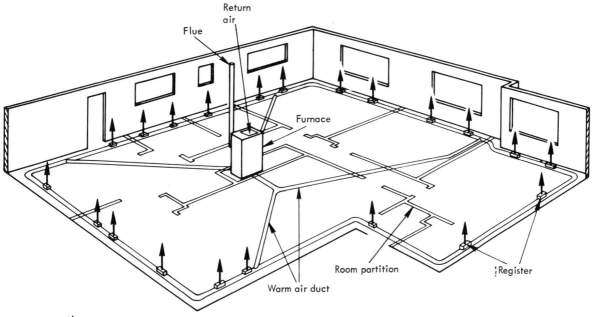

Figure 17.4. Forced air in perimeter-loop system. Return air can be collected by a return register in each room or by a ceiling plenum.

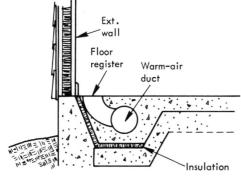

Detail of perimeter duct
(Can be used in any house)

Figure 17.5. Series-loop systems

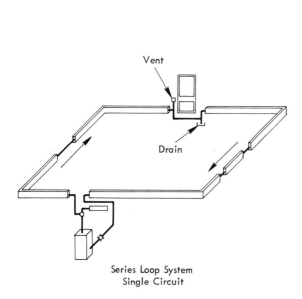

Series Loop System
Single Circuit

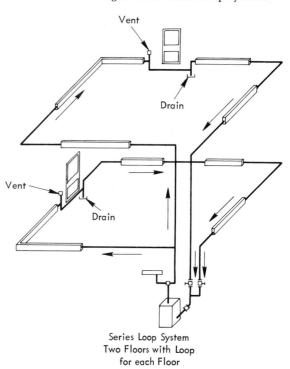

Series Loop System
Two Floors with Loop
for each Floor

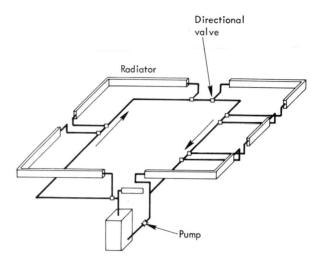

Figure 17.6. One-pipe hot-water system.

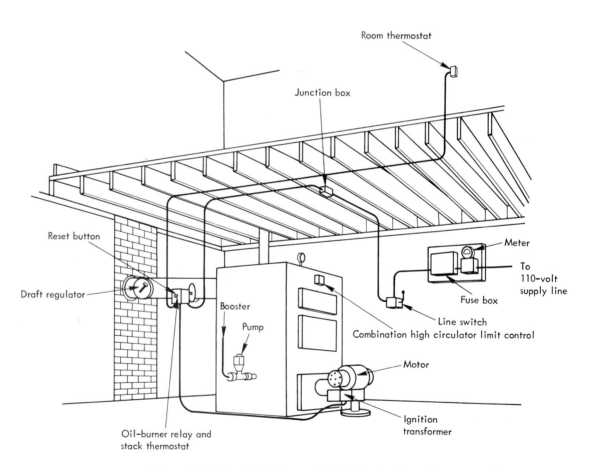

Figure 17.7. Typical oil-burner installation for forced hot water. Note required safety devices. This can also be a gas-fired central plant.

BURNHAM/BASE-RAY
RADIANT HYDRONIC HEATING AT ITS BEST!

RATING DATA

I=B=R STEAM AND HOT WATER RATINGS

FLOW RATE	STEAM RATING		WATER RATINGS BTUH Per Lineal Foot At Average Water Temperatures Indicated						
Lbs./Hr.	Sq. Ft.	BTU/Hr. At 215°F	170°F	180°F	190°F	200°F	210°F	220°F	230°F
2000	3.40	820	550	620	690	750	810	880	940
500	3.40	820	520	590	650	710	770	830	890

NOTES:

1. The Hot Water Ratings at 2000 lb. flow rate are limited to installations where the water flow rate through the Base-Ray is equal to or greater than 2000 lbs. per hour (4 GPM).

2. Where the water flow rate through the Base-Ray is not known, the rating at the standard flow rate of 500 lbs. (1 GPM) per hour must be used.

3. When Base-Ray is equipped with damper, there is no change in the rating when damper is in the open position.

4. Add ½" to length for each bushing.

5. I=B=R Ratings are determined from tests made in accordance with the I=B=R Testing and Rating Code for baseboard type radiation, including an allowance of 15% for heating effect permitted by the Code.

6. Ratings based on active length. Active length same as total length.

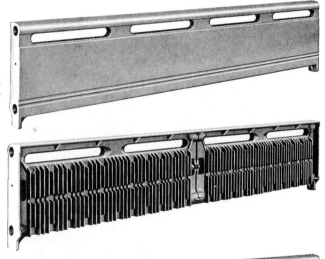

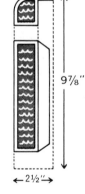

9⅞"

←2½"→

No. 9A

FRONT VIEW

The low trim lines of BASE-RAY blend into a decorative scheme. It's as inconspicuous as floor molding and can be painted to match the color of your walls.

REAR VIEW

For high efficiency and maximum output, fins are cast as an integral part of the BASE-RAY unit. Your guarantee of outstanding performance.

AVAILABLE WITH DAMPER CONTROL

The BASE-RAY damper is available as an optional feature. A flick of the finger enables you to substantially reduce the heat output...important when cooler bedrooms are desired at night. Attractively designed to blend with BASE-RAY, the damper can be quickly and easily installed in new or existing BASE-RAY systems.

Figure 17.8. Base-ray radiant hydronic heating. (Courtesy Burnham Corporation, Hydronics Division, Lancaster, Pa. 17604)

17.2.2   Fuels

Central heating systems can be oil-fired, gas-fired, electrical-resistance-heated, and in some parts of the country stoker-fired with coal. Heating and air conditioning can also be attained by the use of an electrically driven heat pump. Although it is used in all parts of the country, the heat pump is most economical in temperate climates. The numbers shown in Table 17.1 can be used to calculate the cost of the various fuels based on local prices.

TABLE 17.1. Comparative Fuel Prices

(1) Fuel or Energy	(2) Quantity to Supply 1 Therm Usable Heat	(3) Multiply Values in Column (2) by Local Unit Costs	(4) Comparative Costs per Therm[a] of Heat Supplied to Living Space (cents)
Coal	11.8 lb = 0.006 ton	per ton	_____
Electricity	29.3 kWh	per kWh	_____
Fuel oil, No. 2	0.96 gal	per gal	_____
Gas, natural	127 ft^3	per ft^3	_____
Gas, LP (propane)	1.45 gal	per gal	_____

[a]1 therm equals 100,000 Btu. This table can be useful to builder or owner before deciding on source of energy.

17.2.3   Design of a Heating System

It is not the purpose of this section to teach the builder or owner to become a designer of heating systems. The subcontractor who installs the system can design it personally or, as is usual, call on the supplier of the equipment to furnish the proper sizes of furnace, burner, pipes or ducts, radiators or registers, and so on. There are, however, some simple rules for determining the heat loss of a structure (this is for both heating or air conditioning), and this heat loss determines the size of the central plant and its component parts in each space that is to be heated or cooled. Chapter 13 shows how heat loss is determined. The design should be based on the extreme difference between the inside temperature that is required and the outside temperature prevalent in that region. For instance: for heating 70°F inside at 0°F outside; for cooling 80°F inside, 50% humidity at 95°F outside. The individual space design depends on wind direction, sun direction, and exposed perimeter walls. The calculations can be made by an amateur, but such a person is advised to confine himself or herself to calculating only the total load.

17.3   AIR CONDITIONING

Central air conditioning in single-family structures is accomplished by the use of the same blower and ducts that are used for the forced-warm-air system. The circulating air is cooled and thereby dehumidified by a refrigeration compressor or a heat pump on its cooling cycle. The capacity of the system is determined by the heat loss as mentioned in Section 17.2.3.

When the system is being installed, the owner or builder should examine it carefully to see that the joints in the ducts are tight, that there are no sharp bends which impede air flow as well as causing air noise, that ducts going through excessively hot places are insulated, and that the ducts or other parts of the system are properly fastened and insulated from any pipe or structural member that may cause a vibration noise.

17.4 ELECTRICAL

All electrical wiring and equipment anywhere in the country is subject to the provisions of the National Electrical Code and the National Electrical Safety Code. The author does not know of any community which allows anyone but a licensed electrician to install an electrical system in a dwelling.

The builder or owner should check the following items:

That wall and floor outlets are located where they will not be behind furniture and that there are enough of them.

That three-way switches are installed in certain locations for safety and convenience. For instance: turn on stairway light at bottom and turn off at top, or vice versa.

That the service conforms to at least minimum regulations. Most localities now require at least a 100-ampere service.

That there are enough spare terminals in the circuit breaker panel for new future circuits.

That there are sufficient provisions for 230-volt circuits for heavy electrical appliances (stove, window air conditioners).

That all circuits are wired in parallel.

That the electrician labels the circuit breakers.

A wiring strip above a kitchen counter into which appliances can be plugged at any point is a convenience and a sales feature.

All electrical devices should have recognized brand names and be labeled. Copper wire is recommended.

CHAPTER EIGHTEEN

landscaping

18.1 A DEFINITION

The term "landscaping" is used here in its broadest meaning, which is to so arrange the site as to make it most convenient and most attractive for its occupant. This would encompass walks, driveways, drainage, and planting. Chapter 1 examines the various topographical features that must be considered in site selection and Chapter 6 shows how to place a house on a site most advantageously. This chapter will discuss the practical construction methods and materials used to accomplish this.

18.2 SLOPES AND DRAINAGE

It is assumed that the owner or builder has so placed the house that it will require a minimum of drainage and that it has also been provided with foundation drainage, as discussed in Chapter 8. However, it is not always completely possible to avoid all water or to leave the original land contours. There are several simple methods for changing the land contours, as shown in Figure 18.1. The height and depth of all of these solutions must be calculated to fit the slope. The earth bank or earth terrace must be so sloped (angle of repose) as to avoid any possibility of a slide during heavy rains. The stone riprap and the cribbing must be securely anchored into the bank and must provide for drainage. Retaining walls must go below frostlines and must have weep holes for drainage.

The drainage lines must be led to a point that is lower than their elevation so that water pressure may be relieved. If this is not done, the occupant may face a lifetime of pumping.

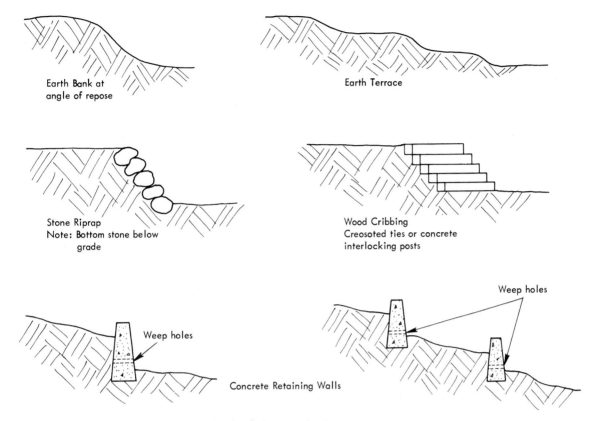

Figure 18.1. Methods of changing land contours.

18.3 WALKS AND PATIOS

This will concern itself only with walkways within the property. A development builder who has to build public sidewalks will have to be guided by the local requirements, which vary widely over the country. Walks within the property can be of brick, flagstone, concrete, blacktop, or almost any wearing surface that meets the owner's fancy. The most popular walks are of flagstone, brick, or concrete. Flagstone and brick for walks are usually laid over a well-tamped sand base. This may cause some heaving during freezing and thawing, but no great damage is done. When this material is used in patios, however, it is recommended that it be laid in a reinforced concrete base. Figure 18.2 shows some patterns for brick and flagstone. When brick is laid for a walk, it should be confined by bricks set upright on either side, as shown in Figure 18.3. Flagstone that is used for walkways should be as large and as thick as possible. The larger the piece, the less it will heave or sink.

Concrete walkways should be at least 3 ft wide and not less than 4 in. thick. To prevent heaving and cracking, it is well to lay the concrete on a well-tamped, thoroughly settled gravel bed. If a walkway slopes more than 5%, it should be broken up by steps or a stepped ramp, as shown in Figure 18.4.

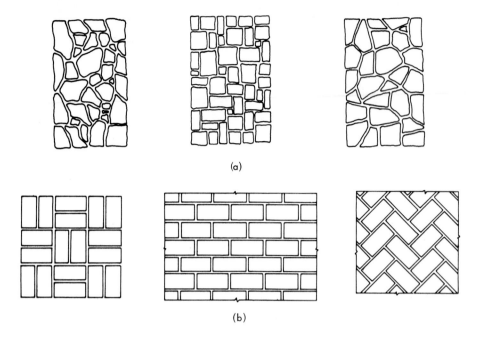

(a)

(b)

Figure 18.2. (a) Typical flagstone patterns.
(b) Typical brick patterns.

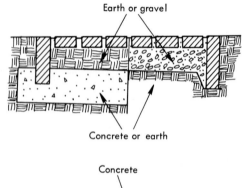

Earth or gravel

Concrete or earth

Figure 18.3. Brick used
to confine walk or terrace.

Figure 18.4. Walkways on slopes.
(U.S. Department of Agriculture)

Concrete

8"

Gravel

Earth

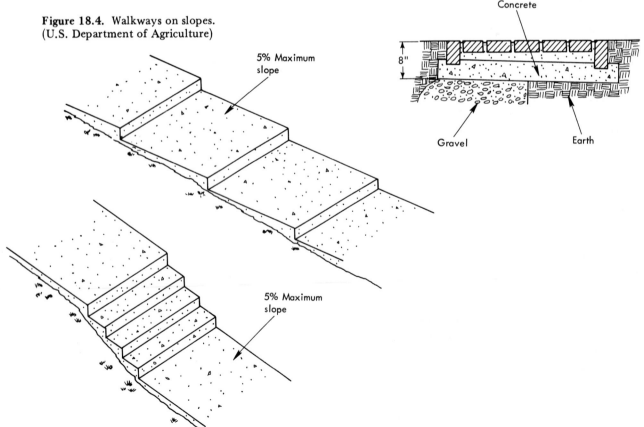

5% Maximum slope

5% Maximum slope

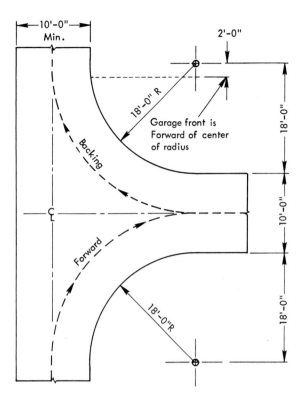

Figure 18.5.
Space required for turnaround.

18.4   DRIVEWAYS

Driveways can be made of concrete or blacktop or they can consist of crushed stone over well-tamped run of bank gravel. The layout of a driveway is most important for the convenience of the occupant. It must be wide enough to allow easy access to carport or garage, and if no space can be allowed for a turn around or backing-in space, as shown in Figure 18.5, it should have enough radius at the curb line to enable a vehicle to back out and turn into the street, as shown in Figure 18.6. If the driveway has over a 7% grade, it is best to pave it or to use some kind of binder on the surface. Loose stone or gravel will wash down the slope and will require constant care. If there is a pronounced slope, it is well to have as much level ground as possible just in front of the carport or garage. Such level ground allows the owner a start if he has to go uphill or enables him to park or at any rate to enter and leave the car space smoothly.

Concrete driveways must be poured over well-compacted gravel and should be at least 5 in. thick. A steel 6 x 6 mesh reinforcement is advisable and does not cost too much. Expansion joints should be used at not more than 10-ft intervals. Blacktop should also be laid over well-compacted gravel and should be at least 2 in. thick. All driveways should be slightly crowned for drainage.

Figure 18.6. Simple street access.

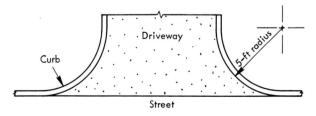

18.5 SITE DEVELOPMENT

The development of the site includes the arrangement of contours, walkways, drainage, and driveways. But there is another important factor and that is concerned with the occupant's life-style.

18.5.1 Planning a Small Area

The following discussion of the planning of a site is an excerpt from a book by the author.* It is addressed more to an owner than to a builder, but it gives the general directions to suit almost any life-style. It is possible that a speculative builder may use some of these directions.

To illustrate the planning process, we will take a small building lot as an example (Figure 18.7). It is quite possible that a cleverly laid out 60-by 100-ft lot can give you as much satisfaction as a 1-acre or larger site. Plan your outdoor space just as you do your space for indoor living and start with a general breakdown of the required space:

A Playground for the Children

This should be placed where it is readily accessible from your back door and where you can get a clear view of it from your kitchen window or from any door or window in your work area. Ten feet square is adequate for a sandbox, slide, and seesaw. You can leave adjacent space for a covered bicycle rack and a small playhouse. This play area need not be seen from anywhere but the back of your house.

A Work Area

This is where you dry your laundry, place your garbage for collection, store your woodpile, barbeque wagon, and lawn mower. Of course, this area can be kept neat, but even so it should not be in public view, and you can tuck it away behind the garage and plant some tall shrubbery around it. It can be as small as 12 feet square and can be paved with loose flagstones, gravel or crushed stone. Do not attempt to grow grass here. It would be a nuisance to cut and it will be trampled constantly. Garbage collection, milk delivery and other service personnel should have free access through this area to your back door.

A Garden Area

The size of this area depends on your gardening enthusiasms. If you would like to grow some vegetables as well as flowers in rows for cutting, you require a minimum of 15 by 20 ft, and it should get the summer sun at least half the day. Be sure that you have a hose connection nearby for watering.

*Buy or Build: The Best House for You (Englewood Cliffs, N.J.: Prentice-Hall, Inc., 1973).

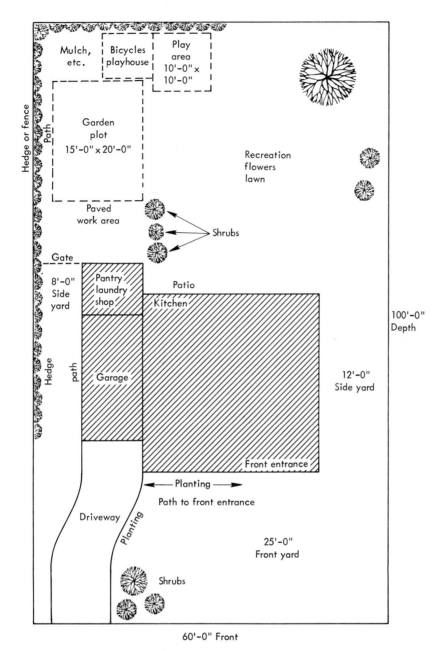

Figure 18.7. Land plan for a small plot.

A Recreational Area

After setting aside space for working area, for play area, and for garden-ing, you have an area over 40 ft square remaining. The number of uses to which you can put this space depends on your own desires. You can build a paved patio adjacent to the back door, which leads from the kitchen, dining room, or living room. This will give you a permanent place for outdoor cooking or space for a movable cooker. If desirable, you can place an awning over part of the area to provide shelter against sun and rain. If you live in a cold climate with a short summer but like to be outdoors as often as possible, you can arrange a sheltered nook which will get the afternoon winter sun and still protect you from the wind. Perhaps you enjoy sunbathing. Arrange a sheltered private spot for this.

The area beyond the patio can be any shape you like. Just remember to avoid corners—curves are much more appealing. This area can contain a small pool surrounded by flower beds. You can encircle your patio with raised planters. You can even install a play area for a sport that requires minimum space.

The Front Yard

If you use a typical zoning code requirement for a lot 60 by 100 ft, you must set the house back 25 ft from the front line (as well as a total of 20 ft from the lines). This 25- by 60-ft area is your front yard, and this is what everybody sees. Passers-by may never see your charming patio or private garden, so they will form their impression of you and the house only from the front.

The first consideration is the location of the garage and driveway. Unfortunately, there is not much that can be done with these on a narrow lot. Perhaps you can curve the driveway slightly so that you break the monotony of a row of houses each with a straight driveway. (A circular driveway, however, can look pretentious, or out of proportion, on all but the largest of lots.) If you curve it, you may be able to plant a clump of shrubs near the front property line. This planting can partially shield the garage from the street and thus prevent a possibly cluttered and never handsome garage interior.

Your driveway can serve as a walk to the front of the house, but in that case you have to provide a footpath from it to the front door. Flagstones, brick, or any other paving material may be used.

Landscaping

Foundation planting around a house requires care. Ask your local nurseryman or plant dealer about evergreen shrubs that do not grow too tall and that require only occasional pruning to keep the house from being submerged. Windowboxes or raised planters around the front door and flower beds around the driveway and walk add immeasurably to the appearance. If you do not want to be bothered with flower beds, you can use perennial flowering shrubs.

Finally, we come to trees. If there are trees already on the property, you can plan around them. If not, you can plant such ornamental trees as holly, dogwood, locust, redbud, flowering crabapple, and such tall-growing shrubs as flowering quince, or forsythia. Try to plant the taller-growing material at the ends of the house in order to frame it. There is a world of possibilities and many good landscaping books to suggest them.

18.5.2 Planning a Large Area

Planning the various areas of activity for a large piece of land is essentially the same as for a smaller one. The difference is that you can allow more space for each activity and you can arrange these areas with relation to the house and the property boundary lines much more freely than you can a small plot.

A large plot of land is also likely to have more variety in elevation and other topographical features. You can, therefore, lay out your areas to take advantage of slopes, large trees, or outcrops of rock. There will be room enough to have a sweeping lawn, a rock garden, or a rose garden.

If your inclination is for a sedentary life, it is possible to keep the natural features of the land and to enjoy them without much gardening or lawn care. In any case, do not feel obligated to use every square foot of your land for something. A piece of meadowland, a grove of trees, or a bare hilltop left in its natural condition can be very appealing.

After all, you are planning your outdoor living to suit your own family's life-style, and the only people you must please are yourselves.

CHAPTER NINETEEN

the residential construction process: a summary

19.1 THE PURPOSE OF THIS CHAPTER

This chapter will summarize on a step-by-step basis the entire process of the construction of a residential building. The preceding chapters have given the details of the various methods and materials that can be used in each trade in approximately the order of their use. This chapter will place them in the order of a progress chart.

19.2 PREPARATION FOR CONSTRUCTION

Prepare a site plan.

Clear the site of brush, trees, and any other obstruction that may interfere with construction.

Obtain a building permit. Have all zoning restrictions been complied with?

Cover the job with insurance.

Have a plumbing subcontractor obtain permits to connect to water and sewer lines. If these are not available, obtain permits for and award contracts for well and septic tank and drain field.

Have an electrical subcontractor arrange for power line.

Arrange for portable sanitary accommodation and, if necessary, a telephone (preferably a pay telephone).

Draw a site plan. Show locations of material storage, truck driveways, miscellaneous excavations, and so on.

19.3 EXCAVATION AND FOUNDATIONS

Scrape top soil and stock pile according to site plan.

Lay out and stake excavation boundaries.

As excavation nears the bottom, notify the local building inspector. All footing and pier hole bottoms must be inspected before concrete is poured.

Arrange for concrete, ready-mixed or job-mixed.

Start footing forms and wall forms for concrete foundation walls or arrange for delivery of concrete block.

Be sure that foundation walls are accessible to a concrete chute. Conveying concrete by buggy is expensive. If concrete blocks are to be used, they should also be chuted down to make them readily available. Concrete blocks are heavy and labor is costly.

Be sure to allow holes in the foundation walls for the entrance of all utilities (water, sewage, electrical, telephone, oil fill, gas, etc.).

Remember hold-down bolts, if called for.

Waterproof the foundation walls and lay drain tile on the footings before backfilling.

19.4 THE STRUCTURE

This section will concentrate on wood-framed structures which can be covered with brick or stone veneer, with stucco, with wood siding or shingles, or with other exterior coverings. Even in the case of solid masonry exterior bearing walls (brick and block, solid brick, block), the interior framing is of wood, and essentially the same schedule can be followed except that the interior framing is now supported by the masonry instead of wood or steel framing.

As the foundation walls near completion, the builder should assemble the lumber and hardware in the order of its use and be sure there are sufficient workers available. The most expeditious erection and closing in a structure is one of the most important things a builder can do. It can save time and money if workers can have shelter during bad weather. Is electricity available for power tools?

If the structure is of wall-bearing light-steel construction, the builder should order the expanded beams, girders, columns, lally columns, lintels, bolts, plates, and so on, many weeks before. When estimating the job, the builder should ask the material dealer how long a lead time is required between order and delivery.

It is safe to start laying sills 3 or 4 days after a block wall or a high-early-strength concrete wall has been completed, if weather and temperature permit. Neither wall should be erected in freezing weather unless mortar or concrete are adequately protected and kept warm.

Do not remove foundation wall concrete forms until at least 2 weeks after pour; more time is better. Backfill after the forms have been removed and the drain tile is in place.

Follow sills with girders and joists and subfloor and start assembling outer walls. Do the same with interior bearing partitions. With wall-bearing masonry walls, start masonry erection.

Do not forget joist bridging. Put in place but do not nail permanently; to be nailed permanently after walls and floors are leveled, plumbed, and braced.

Follow erection of stud exterior walls with second-floor girders and joists. If there is a second floor, follow with stud exterior and interior bearing walls.

Do not forget temporary corner bracing, which should be used as soon as first-floor walls are erected. Use builder's squares, levels, and instrument to keep building level and square as erection proceeds.

Erect roof rafters and prepare for sheathing.

Before sheathing is started, install permanent corner bracing and stud bridging. *Structure must be level and square before sheathing is fastened.*

When masonry-bearing walls are used, the progress of the house framing will depend on the progress of the masonry. This takes more time than the erection of a stud wall and the builder must take account of this. The weather is also much more of a factor than it is for all wood framing.

As framing is nearing completion, start erection of the chimney and fireplace. This must be completed and flashed before roof can be made weathertight.

19.5   ROOFING AND FLASHING

Roofing starts immediately after sheathing is completed.

First sheathing, then flashing, then finish roofing.

Every projection through the roof must be flashed. The builder must supervise this closely. Leaks through defective flashing are very difficult to find.

Finish roofing follows flashing. Closely supervise lapping, edges, overhangs, and proper overlapping of flashing.

The roof is now weathertight.

Gutters and leader follow later.

19.6   EXTERIOR FINISH AND TRIM

Exterior sheathing, which has been protected with a waterproof cover (felt paper, sheathing paper, etc.), is now ready for exterior finish. This can be brick or stone veneer or stucco or metal or wood. The builder must arrange for material to be on the job (and protected from weather) before roofing is completed. (*Note:* Backfill against foundation walls should be completed before this.)

Exterior trim follows exterior siding. The builder should check the door and window flashing. Trim should be predipped or backpainted and be kept perfectly dry before installation. Do not install until after several dry days.

All exterior trim should be completed before interior work is started. Temporary doors are usually used until final finishing, at which time the finish doors are installed.

Leaders and gutters can now be installed.

19.7 PLUMBING, HEATING, AIR CONDITIONING, AND ELECTRICITY

The mechanical and electrical trades will stay with the job from the beginning to the end but are not continuous—coming as they are required. The builder must follow these trades closely to be sure that their work meshes with the general construction.

All these trades are under close supervision of a public authority both as to compliance with codes and compliance with licensing. Special permits are usually required.

As the installation of the mechanical and electrical work proceeds, the builder must be vigilant to see that the trades personnel do not cut through important structural members to make room for their piping or conduits or ducts. The builder should be prepared to patch all such work before it is covered.

19.7.1 Plumbing

The progress chart shows that plumbing starts almost at the beginning of the job and continues to the end, but it is not continuous.

The plumber starts to lay sewer and water lines as the foundation proceeds and can install basement waste, vent, and water lines soon after the framing starts.

The bathtub should be delivered soon after the underfloor is installed but before the interior framing is complete.

There is a wait until the framing is complete before the plumbing contractor starts installation of water lines, wastes, and stacks.

There is another interval until wallboard is completed and the bathroom tiling or wall finish is applied. At this time the plumbing contractor installs the fixtures. Some fixtures, such as the kitchen sink and the bathroom basins, may not go in until just before completion, when the cabinets and counters that receive them are in place.

The contractor must therefore be vigilant to give the subcontractor sufficient notice before each portion of the work so that each subcontractor will fit smoothly into the schedule and will not delay any work that must follow.

19.7.2 Heating and Air Conditioning

If the construction is begun in cold weather, one of the most important things for the contractor to remember is that it is important to get heat into the building as soon as possible. Until the building is heated (to at least 55°F) no wallboard taping, tiling, finish flooring, or interior finish trim can be installed. The heating contractor must be made aware of this when the contract is signed. The heating plant should be on the job before the piping is complete. With its original crating and a heavy tarpaulin to protect it, it should be put in place. The fuel-oil tank, gas connection, or electric connection should be ready while the piping is being installed. The radiators or ducts or wall heaters should be delivered as soon as the house is tight. The finish floor and wallboard must go in before the radiators, but the radiators should follow immediately, and it is possible to do just enough flooring and wallboarding so that radiators or other outlets can be installed and heat can be supplied while the rest of this work is being done.

The heating subcontractor must conceal all piping or ductwork in the walls and floors and must therefore complete all the work before the wallboard can be started. The subcontractor then waits until after the finish flooring and wallboarding to install radiators or outlets for the grilles. Baseboard or recessed radiators should have heavy reflecting foil behind them.

If there is an absolute emergency whereby the construction must stop completely because there is no heat, the contractor can use emergency temporary heat. Almost every building department allows propane-fired space heaters for temporary heating. These are available on a rental basis almost everywhere in cold-weather country. Two or three of them will do the job. They must be watched if they go all night, and this means a watchman for a week or so until the permanent heating system is ready. The contractor should check with the local building inspector and the insurance carrier. Since it is the owner or financial institution that usually carries the fire insurance, the builder must check with either of them also. A written permit from the building inspector and a note from the insurance broker is recommended.

19.7.3   Electricity

The electrician is on the job right at the start to furnish temporary power.

The electrician does not come back again until the structure is securely weathertight. Then all BX or conduit is installed in the walls, floors, and ceilings, where it will be eventually concealed, and then the circuit wiring is run to the panel box. The electrical contractor also furnishes labor and material to connect all mechanical equipment.

The electrician should also have the power company install the permanent electrical feeders and the electric meter. (The meter should, if possible, be located where a meter reader can read it without entering the premises.)

After the wallboarding and finishing are completed, the electrical subcontractor installs the switch and base plates and the lighting fixtures. As for the mechanical work, the contractor should give the electrical subcontractor sufficient notice before each stage of the work.

19.8 THERMAL INSULATION AND SOUNDPROOFING

Thermal insulation can be installed in the exterior walls and between the rafters as soon as the house is weathertight. Care should be taken that it is not disturbed or pushed out of the way by the plumber or electrician or the person doing the sheet-metal work.

Insulation between ceiling joists can wait until the ceiling board is installed and before the upper rough floor is started.

Sound insulation blanketing is also installed at this time.

19.9 INTERIOR FINISH, TRIM, AND FLOORS

19.9.1 Wallboard and Plastering

Wallboard and any plastering should start as soon as the house is weathertight and warm and dry and when all (to be) concealed piping, ducts, wiring, and so on, are in place. Wallboard must be kept perfectly dry. It is suggested that it be delivered on pallets.

Interior finish stairs should be installed. The stairs should be ordered many weeks in advance of installation. In general, ready-built stairs by professional stair builders are superior and will be cheaper than a stair built on the site. This excepts basement or other open-riser stairs.

19.9.2 Floors and Trim

Finish flooring is installed before trim. It must be perfectly dry and installed in a warm, dry house. The floor should then be covered with heavy building paper to prevent denting or scuffing.

Interior moldings, baseboard, chair rail, stools, aprons, and so on, are installed after the flooring, which goes in after the wall board is taped, spackled, and dry.

At this time all interior doors should be hung.

Next, apply for a certificate of occupancy.

The builder is again cautioned about keeping all interior trim perfectly dry before it is installed.

Finish hardware may take weeks between order and delivery. Be sure that enough lead time is allowed.

19.10 FINAL FINISHES

19.10.1 Tiling and Accessories

Ceramic tile, unless it is of a stock size and color, should be ordered in advance; in any case, the builder should make sure that it is available when required. There may also be a problem of obtaining a tilesetter when required. If the builder wants the house completed on time, instructions should be followed carefully. The same caution goes for any tile.

If the shower is precast plastic, it should be installed when the bathtub is installed and should be carefully protected.

The medicine cabinet and bathroom accessories should be on hand to be installed with the tile. The medicine cabinet mirrors and similar breakable fixtures should not be installed until the house is completely secure.

19.10.2   Floor Finishing

The floor can now be sanded and finished and protected. Kitchen and bath resilient tile should be laid.

19.10.3   Painting and Decorating

This is the last step in the completion of a house. Everything to be painted should be warm and dry. Note the cautions in Chapter 16 about exterior painting.

19.10.4   Electrical

Electrician installs lighting fixtures and outlets and switch plates.

19.10.5   Plumbing and Cabinet Work

Kitchen cabinet work should be ordered with sufficient lead time.

Plumbing fixtures are installed after tiling and at same time as kitchen counter work is being done.

19.11   CLEANUP

The builder must keep the job clean during the entire progress and make a final cleanup at completion.

19.12   LANDSCAPING

Landscaping includes finish grading, driveway, walks, planting, and seeding. The amount of work that the builder does depends on the contract.

As soon as the exterior carpentry is completed, the builder should start rough grading (the backfill against the foundation and over oil-tank and pipe trenches should, by now, have had time to settle). Finish grading from the topsoil stock pile should follow. This finish grading may have to wait for better weather.

Foundation planting and seeding depend on the terms of the contract. Driveway and walkways are usually included.

The speculative builder does all of these.

CHAPTER TWENTY

the speculative builder: how to sell the house

Although this subject does not come strictly within the scope of this book, the author has thought that it may be useful to the builder who builds a private residence on speculation.

20.1 SALE BEFORE COMPLETION

The selling of the house before it is completed is, of course, the best way *if the builder has correctly estimated the final cost.* The builder has, of course, estimated the cost of construction. To this must be added administrative costs; insurance and financing costs; profit, which should be over and above the builder's time invested; and a contingency. The amount of the contingency should be based on the state of completion and on the subcontracts that have been let where the builder is sure of a reasonably fixed price. There is also the brokerage commission.

In selling the house before completion the builder has made sure that his or her own capital will be freed and that financial obligations will stop upon completion of the job. The builder is therefore in an excellent position to offer part of this saving as an inducement for a quick sale and a quick closing.

20.2 SALE AFTER COMPLETION

Whether the house is sold before or after completion, it is important that the builder, if at all possible, come to an agreement with the construction lender or another lender to obtain some assurance that the lender will grant a permanent loan to a responsible owner.

When a house is not sold at completion, the builder-owner must estimate what the monthly carrying charges will be as owner of the house and set the price accordingly. Record keeping is very important. The builder should know at any time the price at which he or she can sell the house and still keep a speculative profit.

20.3   ADVERTISING

20.3.1   Signs at the Site

These should go up as soon as the site is being cleared. The sign can say what is to be built, the builder's name and telephone number, and an inducement such as "Early purchase will give buyer choice of finishing materials."

20.3.2   Newspapers

The builder should try for a "punch line" such as "Beautiful Wooded Site" or "Sunny Meadow" or "Traditional Colonial." An ad or two should be run in the Sunday real estate section of a big city newspaper.

20.3.3   Relocation Services

All big companies now maintain a relocation service for transferred employees. These offices will be pleased to list the house. Make sure that there is no brokerage fee involved. Many brokers list themselves as relocation services and charge the builder full commission if they sell the house.

20.3.4   Answering Telephone Calls

No other information should be given on the telephone than is on the sign. The callers should be asked if they are brokers. The information regarding price, terms, and so on, should only be given when the builder meets the prospective purchaser on the site. The builder must make sure that the person or persons who come to view the house are the principals and are not representing a third party. Legal advice is suggested regarding this matter.

20.4   LISTING WITH BROKERS

The builder must determine when to enlist the services of a broker to sell the house. There should be a clear understanding of the obligations between the builder and the broker.

Brokers have cross listings with other brokers, but only one commission will be paid.

The price must be firmly established.

The terms and conditions of the sale must be established.

Multiple listing is suggested.

The key to the house must be given on a strictly limited basis.

The builder must state in *writing* that the house is for sale "as is—where is" or what additional work could be arranged for. Putting the terms in writing is important. Brokers may promise things in order to make a sale.

index

Moldings, 233, 237, 238 (fig.)
Mortar, brick, 174, 175 (table)

Nailing schedule, 122 (table)
Nails, 121 (fig.), 129

Occupational Safety and Health
    Act (OSHA), 65, 92
Odd-shaped lots, 19, 20 (fig.)
Oil-burner installation, 270, 272
    (fig.)
One-pipe hot-water system, 270,
    272 (fig.)
Overhead:
    builder's, 59
    general, 86
    job, 86
    owner-builder's, 60

Painting:
    costs, 58
    exterior, 254, 255 (fig.), 256-59
    interior, 207-8, 260, 261 (fig.),
        262 (fig.)
    specifications for, 52-53
Panel siding, 167 (fig.), 169-70
Parquet flooring, 210
Partition framing, 145
Patios, 277, 281-82
Pattern bonding, of brick, 174
Payment schedules, 68-69
Permits, pre-construction, 72
Personal loan, 70
Piping, 218, 265
Pitched roofs, finishing of, 186-87
Plans, house (*see* House plans)
Planting, foundation, 282
Plasterboard, 198
Plastering, 201, 202-5, 289
Plastic, water pipes of, 265
Platform frame, 129-32, 142
Playground, planning for, 280
Plumbing:
    in construction schedule, 287
    costs, 58-59
    materials, 265-69

Plumbing (*cont.*):
    specifications, 32, 53-54, 265
    symbols used in, 43 (fig.)
Plumbing, of a wall, 138-39
Plywood, 158-62
    for exterior cladding, 176-77
    for interior finish, 205
    for sheathing, 164, 165 (fig.)
    for subflooring, 135
Polystyrene insulation, 214 (table)
Post-and-beam framing, 151-53
Preservatives, for wood, 255
Productivity, labor, 158
Progress schedule, 82-85

Quality checks, 87
Quantity checks, 87
Quantity takeoff, 154-57

Rafters, 149-50, 157
Record keeping, by contractors,
    66-67
Recreational area, planning for,
    281-82
Reflective insulation, 216-17
Reinforcement, foundation, 115
Resilient flooring, 209, 210
Rock wool insulation, 214 (table)
Roller painting, 256-57
Roofs:
    in construction schedule, 286
    costs, 58
    finishing, 186-89
    framing, 145-51
    insulation, 217
    types of covering, 179-86
R values:
    defined, 213
    for insulating materials, 214
        (table)

Saddle, 194 (fig.), 195
Safety, on-the-job, 91-92
Sanding, of floors, 260
Schools, as site selection factor,
    3-4, 8